Courtesy, Larry Oliver, Woodland, Calif.

This scene on the range symbolizes both the romantic and the economic sides of the cattle industry. The cow's affection for her calf, signified by her following a stranger, and the cowboy's willingness to work in any kind of weather to save the calf's life and ultimately benefit his pocketbook, are typified in the wood carving by the cowboy artist Larry Oliver.

APPROVED PRACTICES IN BEEF CATTLE PRODUCTION

by

ELWOOD M. JUERGENSON, Ph.D
Formerly, Teacher Educator in Agriculture
University of California
Davis

THE INTERSTATE
Printers & Publishers, Inc.

Danville, Illinois

Library of Congress Catalog Card No. 79-88958

ISBN 0-8134-2093-8

PREFACE

The purpose of this book is to furnish a comprehensive list of activities which involve approved practices, along with information on how they should be conducted in the beef enterprise. An approved practice is one which has been tried and tested by state agricultural experiment stations, the U.S.D.A., and/or successful farmers in the community and found to be a desirable practice to follow for most efficient beef cattle production. Consequently, it is important that the producer know what the approved practices are and how to carry them out in order to attain the greatest success in the beef business. A summarizing list of suggested approved practices is presented in Appendix C.

The primary aim of this book has been to include all of the more important activities involving approved practices in the beef enterprise. The information has been carefully selected and condensed to eliminate a vast amount of reading in order to determine the best methods and procedures to follow in carrying out the practices.

A considerable amount of information has been obtained from state agricultural colleges, experiment stations, and the U.S.D.A., as well as from discussions with livestock specialists and successful farmers. This should provide the reader with facts concerning the latest techniques which have been tested and proved valuable. A number of plans and diagrams of beef equipment have been included, together with sources from which more extensive plans can be obtained should the reader desire to construct the equipment.

One of the author's chief purposes in presenting this material has been to tell *how* each activity should be done in order to accomplish the approved practices involved. Whenever possible, specific details have been given. Many activities and approved practices common in beef production can be used as described on any ranch or in any area. Some, however, may require minor local adaptations.

This book should be especially helpful to those individuals who desire to enter the beef cattle industry, in addition to farmers (full-time, part-time, or suburban), teachers and students of agriculture, FFA members, 4-H Club members, county agents, and other persons interested in efficient beef production.

The Author

ACKNOWLEDGEMENTS

The author wishes to express his appreciation to those who furnished photographs for use in this publication. He also wishes to express his appreciation to all those who granted permission to have reproductions made from their plans of beef equipment and statements quoted from their publications. Credit has been given for the use of these materials as they appear in the book. The presentation of much of this valuable material would not have been possible if it had not been for the excellent cooperation of all those who granted permission to use their materials.

Special acknowledgement is given to the breed associations, commercial companies, the U.S.D.A., and the following state colleges and universities for quotations used:

Arizona	Kansas	Oklahoma
California	Michigan	Oregon
Colorado	Missouri	Pennsylvania
Florida	Nebraska	South Dakota
Georgia	New York	Tennessee
Iowa	Ohio	Texas

The author is especially indebted to Dr. Floyd Carroll of the Animal Husbandry Division of the College of Agriculture, Davis, California, for reading the chapter on selecting beef and to Reuben Albaugh and Dr. Glen Spurlock, Livestock Specialists, University of California, Davis, for their counsel and many suggestions.

TABLE OF CONTENTS

Chapter I

OPPORTUNITIES IN BEEF PRODUCTION

The world is beginning to learn with increasing rapidity there is no cheap source of food. This is especially so for the best of all foods—meat, and particularly beef. Worldwide inflation and population increase are the main reasons. Another is that times are good, and when people are prosperous they eat more meat, especially beef. Many myths exist relative to profits, yet few consumers have any idea of the cost of raising, processing, and merchandising food. However, they are beginning to learn as the price of food increases. In the United States, consumers spent a world low of some 16 per cent of their income to bring home the groceries, compared with a high of 70 per cent for some African nations and an average of 62 per cent in western and eastern European countries, according to American Beef Association figures. In spite of this, beef is selling for a bargain when compared to increased farm costs or average wages of today's workers. Increased food costs will be met by tremendous social and political pressures, and meat seems to be the focus of the attack. For these reasons beef producers must be extremely alert and well informed in order to weather conditions which could reduce or eliminate

Courtesy, Union Pacific Railroad

Fig. 1.0—An intermountain cattle scene. Grass, cowboys, and moving cattle have long symbolized the beef industry. While this is a picturesque business, easy profits are not possible on such an operation but demand strict attention to all phases of production.

profit and curtail the mainstay of free enterprise in American industry.

Since the beginning of civilization, the ruling people and perhaps the most progressive have been meat eaters. Beef, cooked in one way or another, forms the basis around which most of our meals are prepared.

These two conceptions go hand in hand, for beef along with other meats is a basic food. Not only is it

Courtesy, New Holland Co.

Fig. 1.1—The production of an adequate supply of high quality, economical feed is the first approved practice in successful beef production.

one of the most palatable foods but also provides body building protein, is rich in vitamins and minerals, and supplies energy for work that we do. The production of beef, therefore, is an essential industry from the standpoint of the average citizen.

To these basic concepts must be added the lure and romance of agriculture in drawing people into the industry, especially at the production level. The result is not surprising, for present are many natural instincts, including love of living things, wonder at the emergence of a new crop in spring, the mystery of birth, the challenge of competition—all heralding a way of life in the out of doors in rural America. The cattle industry perhaps more than any other phase of agriculture exemplifies these instincts and offers the independence characteristic and synonymous with the cowboy, the cattleman and ranch life.

Therefore engaged in this great industry are found not only the commercial stockmen and feedlot opera-

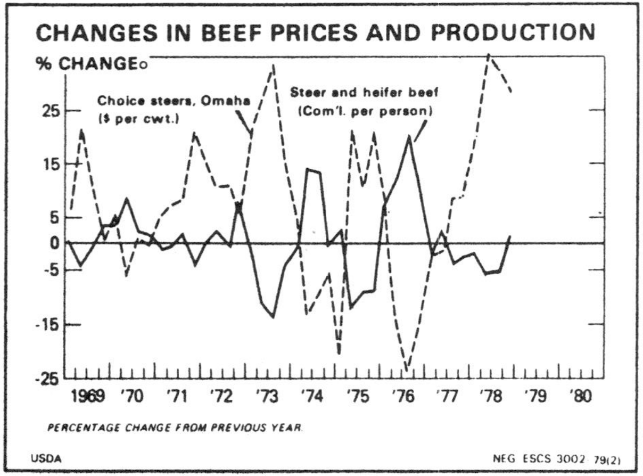

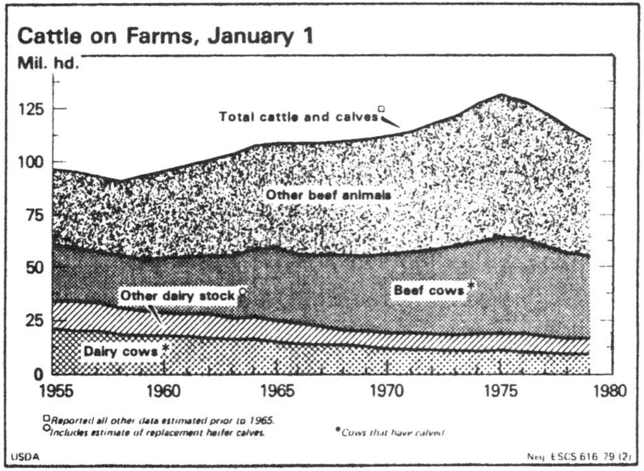

Fig. 1.2—The long range outlook for beef is good. Increasing population and expanding worldwide consumption are favorable aspects. However, beef prices are erratic and cattle numbers cyclical, as the charts above indicate. Note that cattle numbers in 1979 are near that of 1970. During periods of high prices, most successful livestock men will cull rigidly.

tors who must make a profit, but those who enter simply to be associated with an enterprise for its esthetic advantages.

Cattle raising is also necessary for permanent agriculture. When crops are fed to animals, about 80 per cent of the fertilizing element of the feed is recovered in the manure. This factor is of great importance in maintaining soil fertility.

Low Labor Requirements

Labor requirements for beef are only 20 to 40 man-hours per cow per year compared to 100 to 150 hours for milk cows. Self-feeding from trench or horizontal silos and other special methods may be used to further reduce labor requirements. This low labor requirement is due largely to the fact that with beef the calves milk

CATTLE AND BEEF PROJECTIONS—1980 (48 STATES)

Supply	1971	1980	Percent Change
January 1, Cows (1,000 head)	49,841.0	56,300	+13
Ratio of Calf Crop to January 1, Cows	94.1	95[1]	
Total Calf Crop (1,000 head)	46,896	53,500	+14
Cattle Imports (1,000 head)	875	600	
Non-Fed Steer and Heifer Slaughter (1,000 head)	2,034	—	
Fed Cattle Slaughter (1,000 head)	26,543[2]	38,510	+45
Calf Slaughter (1,000 head)	3,821	1,400	−63
Cattle Slaughter (1,000 head)	35,896	46,900[3]	+31
Beef Production (million pounds)	21,870.0	29,700	+36

[1]Same percentage as 1972.
[2]Fed cattle marketings for 23 states plus 5 per cent.
[3]Allowing for an increase in inventory of one million head during 1980.
Source: University of California, Davis AXT-381.

Courtesy, Charles Adamson

Fig. 1.3—An oxcart in Vietnam. While in the United States present day cattle are used primarily for meat or milk, in many parts of the world they are also used for draft purposes. This pair of animals is pulling a load of farm produce to market but on the following day might be hitched to a plow.

the cows. This fact tends to strengthen beef as a part-time farming enterprise. It is also an inducement for the man who wishes to use beef as a supplement to a major farm enterprise.

Low labor requirements aid the farmer who must compete with the higher wages and shorter hours of the industrial worker.

Wasteland and waste crops, semi-arid land or rough land unsuited to intensive farming are put to work producing food for people by beef cattle.

Not only do cattle convert crops into products of greater value, but also facilitate proper management and marketing of farm products. Each carload of fed cattle that goes to market carries with it one carload of hay and one of grain thereby eliminating expensive freight costs for two crops.

No other agricultural pursuit offers greater opportunity to achieve personal satisfaction and financial reward than does livestock production. Beef cattle raising is high on this list, for it is necessary for a nation's health and welfare, basic for sound, permanent agriculture, and a happy way of life for many rural people.

Fortunately for us, total cattle numbers, both beef and dairy, have increased with the growth of population in the United States. The ratio of cattle to people has varied from 50 to 62 cattle per 100 people during the past 25 years in relatively smooth 10-year cycles.

History would indicate that this ratio of cattle to people will continue to decline. *The ratio of cattle to people has been declining* since 1953, even though total cattle numbers have continued to rise.

Consumer Demand Increasing

Consumer demand presents a most optimistic outlook for beef cattle, which has many implications for the beef producer. Here's what's happened and what appears likely for the future, according to *Iowa Farm Science,* Vol. 22, No. 7:

> In the past 25 years, per capita consumption of beef has nearly doubled.
>
> In the past, rising income has meant increases in beef consumption—and most indicators point to further rises in income during the years ahead.
>
> More people are eating meals away from home—and beef is a big item on restaurant menus.
>
> A greater per cent of the total population is in its "teens," and teenagers are big consumers of hamburger.
>
> And, while national trends are not absolute, human population appears to be increasing.

All these factors point to a strong demand for beef in the United States.

The growing demand for beef in the past has been

met only in part by an increase in the number of cattle. Part of the increased beef production came from the increase in the amount of fed beef. Today, 65 to 70 per cent of the federally inspected slaughter beef is fed.

More animals that once were slaughtered as 900-pound grass fats are now 1,100-pound fed animals which are produced more rapidly than grass fats. Another factor in the increased beef production is the larger number of dairy calves that were once slaughtered for veal but that now are fed to maturity.

Future increases in beef production may not be gained so easily. Some increased production can be obtained through increased production efficiency in beef cattle herds. The rest of the increase likely will have to come from increased cow numbers.

The greatest opportunities for increasing production efficiency appear to be improving the calving percentage and weaning heavier calves.

Opportunities in the business of producing beef are numerous and varied, but all depend upon one factor—an economical source of feed. Unless the cattle raiser has an economical supply of feed, there is little chance for continued success.

Beef production, depending upon facilities, divides itself into a number of rather well-defined categories.

Opportunities in Beef Production

1. Producing purebred beef cattle.
2. Producing cattle on the range.
3. Producing beef in the feed lot.
4. Producing beef from by-products.
5. Producing cattle on permanent pastures.
6. Producing beef in drylot cow-calf operations.
7. Producing beef for home consumption.
8. Fitting beef into part-time farming.
9. Starting in the beef business.

Courtesy, William Krueger

Fig. 1.4—Indian type cattle in Liberia, Africa. Cattle from India have been crossed with the Ndama native type cattle of Liberia to produce animals like those shown above. Such animals are adapted to the environment and do well, but unfortunately are seldom eaten. This illustrates another use of cattle for here these animals are used to show status or as a symbol of wealth for the owner and are seldom slaughtered even though the population may be on a protein-deficient diet.

1. Producing Purebred Beef Cattle

The producer of purebred cattle has two main functions in the livestock industry. First, to improve the breed so as to make faster, more economical gains at the same time satisfying market demands. Second, to serve as a source of good cattle to grade up and raise the general efficiency of commercial herds.

In order to successfully accomplish the preceding, the breeder of registered cattle must show somewhat greater skill in breeding, feeding, management, advertising, and salesmanship than does the commercial cattleman. For these reasons, it is a specialized part of the livestock industry and requires most investment per head in cattle and equipment. Personal qualifica-

Courtesy, Soil Conservation Service, U.S.D.A.

Fig. 1.5—Purebred beef cattle grazing alfalfa and orchard grass pasture in Kentucky.

tions of the breeder are important because he must have patience and ability to withstand occasional disappointments.

There are ample rewards, however, for those who qualify. With cattle, the business is far from overcrowded since, according to American Hereford Association, only 3 per cent of our cattle population are registered purebreds. Therefore, good purebred beef animals will always bring a premium price.

Young people in the various agricultural clubs have an excellent opportunity to enter this field, as they can start small, profit from their mistakes with a fewer number of animals, and develop their reputation over a long period of years.

With the exception of a few wealthy individuals who

Courtesy, Union Pacific Railroad

Fig. 1.6—A fine bunch of feeder cattle from a western range going to market or feeding lot.

"buy in," most breeders find it profitable to develop a commercial herd first. After they understand the beef business, a few purebred animals are obtained and the purebred herd is gradually developed to the place where it will sustain itself.

2. Producing Cattle on the Range

Range cattle generally refers to cattle produced on native forage in areas unsuited to general farming.

This would include rough and hilly land, untillable land, the great plains area, forests and public domain, low areas subject to overflow, and swampland. Obviously any area capable of raising grass and forage for cattle and unprofitable for general farming can be termed range. Almost every conceivable variation in soil, climate, and topography is present in the areas mentioned. In addition, there is probably no place in the nation where range conditions do not exist; however, the vast acreage of range cattle country is in the western half of the United States. While many purebred herds are located in range country, most of the cattle produced here are from commercial herds and constitute a great source of feeder cattle.

Premium prices do not exist for producers of range cattle, and the stockman must depend entirely upon his ability to manage his herd so his livestock can convert grass to meat in the most economical manner.

Some cattle are sold as "grass fat" off the ranges, but many of them go to the feed lot to be finished on grain, oftentimes hundreds of miles away from where they were dropped as calves.

The practice of creep feeding calves while they are nursing their mothers results in heavier, well-developed calves at weaning time. These are ready for market without further finishing although they often continue to be fed until they reach the size of "baby beef." However, at present, creep feeding is not a common practice on the range.

In the past, profit was possible by sheer magnitude of operations. While "cattle empires" still exist, success in the range business now depends on factors other than size.

Conservation of feed, reseeding with improved grasses, proper management, and rotation of pastures, along with the right breeding and marketing program,

Courtesy, Soil Conservation Service, U.S.D.A.

Fig. 1.7—Cattle grazing a western range. This is a typical cow-calf operation.

make it possible to succeed with a much smaller number of head. However, an economical unit generally contains several hundred head of stock.

Persons entering this business must first obtain a range capable of carrying enough cattle to fully utilize their time and equipment. Good native cows are suitable for the foundation stock. These can be purchased reasonably from local sources. Then, by use of excellent type bulls, the herd can be upgraded and constantly improved.

The commercial cattleman who sells his stock at market price can best increase his profit by lowering the cost of production. This he can do by improving his range so that it requires fewer acres to the animal and by improving his cattle so they will better utilize feed and command a higher market price when sold.

The range of operations that can take place on a single foothill ranch can be seen by examining the systems that were set up to study income possibilities.

The California Agricultural Experiment Station discusses the six plans that were examined.

Plan 1: A beef cow herd, calves sold as weaners.
Plan 2: A beef cow herd, calves sold as yearlings.
Plan 3: An all-stocker operation, with 500-pound stockers bought in October.
Plan 4: An all-stocker operation, with 550-pound stockers bought in January.
Plan 5: A beef cow herd, calves sold as slaughter steers and heifers.
Plan 6: An all-stocker operation, with 500-pound stockers bought in October and sold as slaughter steers.

Many small farms have land areas unsuitable for farming, and can profit by keeping a few head of range beef cattle to harvest crops that would otherwise go to waste. In addition to these herds, cattle that are used to clean-up crop residue such as beet tops, grain stubble, cornfields, or any feeds left after harvesting of the main crop are considered to be range cattle.

3. Producing Beef in the Feed Lot

Cattle in the feed lot generally refers to animals that are being fed harvested crops so as to put them in good marketing condition. These cattle on the whole are around 650 to 700 pounds in weight when first put on feed, but are not especially fat. They are fed for varying lengths of time and are called short fed (60-90 days) or long fed (120-180 days), although they may be kept longer if the operator has excess feed on hand and thinks the market will demand his kind of cattle. Animals placed on feed at this weight will finish out around 1000 pounds live weight which will yield a desirable carcass for present day market standards.

Obviously most of the cattle feeders do not raise their own cattle but purchase them from the range country. The large feeder markets lie in between the western ranges and Corn Belt, although important feeder markets exist in the East (notably Pennsylvania) and some even on the Pacific Coast.

Most of the feeder cattle are purchased from September to the end of December when the movement of cattle from western range country coincides with the grain harvests throughout the nation.

While there are many large operators, most farms in the Midwest purchase about two carloads per year to fatten out on their crops. Year in and year out cattle feeders believe they get more money back from fattening cattle than if they had sold the grain outright. Feed-lot operators make their profits from two main

Courtesy, Union Pacific Railroad

Fig. 1.8—Fattening cattle in the feed lot. Note the method by which feed is handled via the bulk auger from the truck. Feed mixes are the most economical possible and are often fortified with vitamins. Continuous observation of cattle is maintained in order to keep animals "on feed" and lower the death loss.

Fig. 1.9—Cattle in a custom feedlot. Stretched cable is used in order to allow for wind circulation so animals keep cool in summer. Many producers now send their animals through a custom feedlot and thus market them as finished cattle.

sources: (1) the increase in weight of the animals, and (2) the increase in market grade, so they make a profit on the increased price per pound even on the original weight of the animal when first purchased. This latter is not always true, but gives rise to such terms as "margin," which means the difference between cost per pound of animals and the selling price per pound. An additional profit is often made by allowing hogs to "follow" the cattle to clean up waste and undigested grain.

In recent years a specialty in the beef feeding industry known as "Custom Feed Lots" has come into prominence. These feed lots are found in the major feed producing areas and furnish a place for ranchers to fatten their cattle where no home facilities exist. They provide pens, labor, and other facilities needed to handle cattle, and in addition have large-scale mills and concentrate facilities to supply and feed great numbers of cattle.

Under these conditions, if a rancher has feeder cattle

Courtesy, Pacific Molasses Co.

Fig. 1.10—Molasses sprayed on unpalatable, cheap grass will make good feed. This is a desirable practice in utilizing a by-product of the sugar beet industry.

and he does not wish to sell them as such, he can fatten his own cattle in a custom feed lot.

There are other reasons why grain and concentrate producers like to fatten cattle. First, livestock is an easy way to market the large tonnage of crops tilled from the soil. Freight costs are therefore at a minimum. Another very important factor is that fertility is maintained by leaving the bulk of the fertilizing elements on the farm. This is especially true if the manure produced is handled properly. The weed problem is also helped whenever cattle graze fence rows, etc. Finally, a real contribution is made to the management program in getting better utilization of labor throughout the year and balancing of farm operations.

Feed lot cattle constitute one of the major types of cattle raising in the Midwest and large concentrate producing areas, but even on the Pacific Coast, about 15 per cent of the industry is confined to feed lot production.

Poor judgment can be disastrous to the cattle feeder. Success in this type of operation depends first upon having the facilities to produce concentrates such as corn and barley. In addition, the managerial ability to purchase and sell cattle wisely as well as cattle feeding know-how are required. As is oftentimes said, "The eye of the master fattens his cattle."

4. Producing Beef from By-Products

The first requisites for success in the beef cattle enterprise are sufficient range or pasture together with land to supply feed during the winter or non-growing season, adequate equipment, and sufficient water.

Beef cattle produced from by-products have the same requirements as those on any farm or ranch except that the main source of feed is the plant by-products of some generally large-scale commercial operation.

Among the earliest and best known by-product feed lots are the beef feeding sheds surrounding the sugar beet processing plants scattered throughout the western and intermountain states. Here the principal feed is wet beet pulp. However, this is by no means the only by-product useful in feeding and fattening cattle. Dried fruit, cottonseed meal, and other protein rich plants have for many years been standard feeds for cattle. Of late years, fruit pulp, especially citrus pulp, has played a leading role in the beef industry in Florida and California. Still other products, molasses, brewery and distillery by-products, constitute an important source of feed. Out on the farm, residue from crops such as beet tops or grain stubble is almost always used to feed and fatten cattle.

Practically all of the feed lots fattening cattle from by-products are large-scale operations. Often they run thousands of head of cattle in a relatively small space.

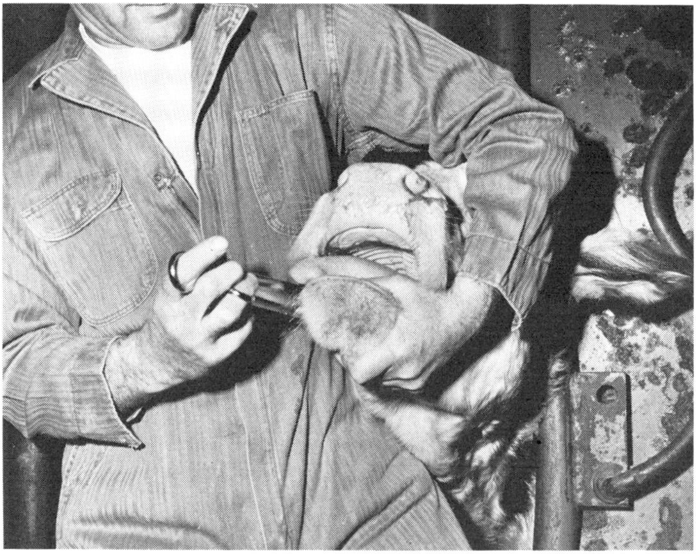

Courtesy, Merck & Co., Inc.

Fig. 1.11—Producing beef on by-product feedlots is a large operation and is mechanized as completely as possible. However, one operation that is still on an individual basis is that involving the health of the animals. Most feedlots employ a full-time veterinarian. Here one is showing the proper way to dose an animal for parasite control or for other reasons.

These concerns, like the small operator, purchase their cattle as feeders and feed them out until the stock reaches the desired weight and finish.

Regardless of what the main feed or by-product is, it must be supplemented with some roughage and additional concentrates, generally corn or barley. Most of the by-products can be fed free choice to the cattle under which conditions they consume large amounts of that feed. However, some notable fruit by-products, such as raisins or prunes, must be regulated because of their laxative effect.

One of the factors that characterizes large-scale operators, feeding by-products is the complete mechanization of the plant. Corrals, feeding troughs, runways, etc., are so arranged that hand labor is at a minimum and feeding can be done quickly and easily by use of mechanical conveyors, trucks, and other devices.

The feeding of by-products is a very commendable asset to the cattle industry because waste products are salvaged to benefit mankind. Many agriculturalists are of the opinion that large amounts of roughage, such as grain stubble, go to waste each year that should be converted into meat by livestock. Perhaps the soundest investment farmers can make is to inventory their operations and see whether or not there are waste crops on their own farms that could be converted into beef.

5. Producing Cattle on Permanent Pastures

Growing grass has often been called "green gold." This is a title well deserved, but only true when livestock convert grass into products useful to man.

Permanent pasture is a relatively new development in the beef cattle industry and is proving to be one of the real agricultural discoveries of the present day. In general, it refers to tillable land planted to a mixture of perennial plants that are nutritious and palatable to livestock. Most of the species of plants used are legumes or grasses, although a few other species may be utilized. The varieties planted in the pasture are those that are best adapted to a particular area and will in addition give nutritious grazing over a long period of the year. Where summer rainfall is plentiful, no irrigation is needed, but in arid regions or where summer rains are lacking, water must be supplied. Such fields are termed "irrigated pastures." Permanent

Courtesy, Soil Conservation Service, U.S.D.A.

Fig. 1.12—Beef cattle grazing irrigated pasture on Walking "T" Ranch, Ellensburg, Washington.

pastures, unlike native range, must be cared for just like any other cultivated crop.

However, the extra effort is well rewarded. Reports of one acre supporting one mature animal are not unusual in comparison to ten, twenty, or even many more acres that may be required to support one animal under range conditions. For these reasons, fields in permanent pasture are generally small and well fenced.

The large amount of feed that can be obtained from permanent pasture is, of course, its most important feature. Many other advantages exist. Cattle harvest their own food by grazing, thereby saving labor. The forage obtained is highly nutritious because it is green and growing, and generally contains a high percentage of protein rich plants. Land that tends to erode can be held in place by the sod. Fertility is improved not only by nitrogen returned to the soil by legumes, but droppings are distributed evenly over the field.

Stockmen can succeed by utilizing permanent pasture with a much smaller number of acres than in any other type of beef raising.

There are disadvantages too. Permanent pastures

can be overgrazed like any other crop. Weed problems soon appear. Bloat is also a real danger and constant watch must be exercised. Parasite problems increase. If irrigation has to be practiced, extra labor and expense is involved.

In spite of these disadvantages, permanent pastures are making a real contribution to the beef industry. Furthermore, in many cases they offer an opportunity for many farmers and others to become established in the beef industry with less capital investment than any other method. In addition, land that was lost to agriculture through cultivation is being rehabilitated by planting it to permanent pasture for grazing with beef.

6. Producing Beef in Drylot Cow-Calf Operations

Typically and historically, the beef industry has been associated with open space, the range, or at least irrigated pastures. Closer confinement or feedlot operations are a relatively recent focus of the industry and when used were generally part of the fattening and finishing process. As space and rangelands become more of a premium, attention is centered on more intensive operations and a cow-calf setup on a small acreage becomes a distinct possibility. In fact, the University of Arizona reports it has kept a sizeable beef cow herd on half an acre for 11 straight years. Its stocking rate was up to 45 head in each pen of 16,000 square feet, or about 355 square feet per cow with calf. Cattle do well on drylot, but a cheap source of feed is necessary, as it takes a minimum of 3½ tons of hay (e.g., baled alfalfa) to keep a cow one year. It is important to have adequate bunk space so all cows can eat at once.

Some factors which researchers found important in confining cattle on drylot were:

1. Artificial insemination was more convenient and practical.
2. *Calves* should be given special attention, including navels treated with iodine, vitamin A injected, and good records kept on each calf.
3. Calves should be creep fed.
4. Vitamins are given twice a month to each cow by mixing 500,000 I.U. of vitamin A in feed.

Some cattlemen will probably find it convenient to combine drylot operations with other methods.

7. Producing Beef for Home Consumption

Since the advent of locker plants and successful home freezers, beef production for home use has become an important consideration in the farm plan. Of course, on a cattle ranch or livestock setup operating primarily for the production of beef, it is a simple matter to select a fat animal to slaughter as the need arises. On the other hand, many farmers grow crops other than cattle for their livelihood or are employed by industry and live on small farms. In order to supplement their income or assist the family larder, they grow much of the food used by the family. Under these conditions, one or two beeves may be produced per year from a variety of home-grown roughages supplemented with small purchases of concentrates.

Many of the animals used for home consumption have some dairy breeding in them as a result of the family milk cow being bred to a beef bull. This is a very satisfactory arrangement which supplies both meat and milk economically to the home. Dual purpose cattle are admirably suited for such an arrangement, as the dams supply milk for family needs and their calves make very acceptable beef animals. Other farmers simply buy one or two feeder calves from central mar-

Fig. 1.13—Buy good heifer calves or yearlings like these as one way of entering the beef business.

kets to grow out and fatten. Oftentimes people are quite exacting about the quality of meat they eat and take great pride in conditioning animals to just the right degree for their own table.

Because of the fact that filling the home locker is more important than producing the maximum number of pounds of beef, most of the cattle slaughtered will weigh less (about 600-800 pounds) than cattle fed out to sell.

On farms that have pastures for livestock or fields for other crops, it is a common sight to see one or two beef animals grazing along with other stock as an economical way of producing meat for the table. Even in orchards, an acre or so may profitably be set aside and planted to irrigated pasture to produce beef and other animal products for the family. Leaflet 211, California Agricultural Experiment Station, Davis, has considerable information regarding small-scale beef raising.

8. Fitting Beef into Part-Time Farming

Part-time farmers, or those having jobs in industry on a 40-hour week, often find time to raise cattle on a small scale. Low labor requirements favor this. Some of these people raise beef as a hobby to show at fairs. This type of farm will probably become more numerous because of the increasing popularity of homes in rural areas. Meat for the family and healthful outdoor work are reasons enough for part-time farmers to consider beef production. The recreational factors of this type of farm should not be overlooked. Further development of through-ways and the tendency for industry to move out of thickly settled areas will tend to increase part-time farm ownership.

9. Starting in the Beef Business

Build Your Own Herd

If you want to build up a herd, you can obtain stock in several ways.

Buy heifer calves or yearlings—These are usually available in the fall from western ranges. You can get them at a central market or from a commission company. Western animals are usually well bred. They are of uniformly good beef-type, well quartered with wide deep bodies. Such heifers, particularly calves, are open and usually disease free.

The disadvantage of buying young animals is that there is no income for nearly two years before they produce their first calves.

Buy old range cows with calves at foot and re-bred—Frequently a commission company can obtain these for you on order. These cows often have several years of usefulness ahead of them. The rancher may

Fig. 1.14—Range cows with calves at side make good animals with which to start a herd. Frequently the cows are rebred, and resulting calves are generally from good bulls. Occasionally old range cows culled because of age may be good buys for the beginner who can provide better than average care for a season or two.

sell them for one of many reasons. He may not have enough forage, his herd may have to range over a large area not suitable for older animals, or he may be replacing range cows with younger females raised in the herd.

Be sure to have these cows tested for tuberculosis and brucellosis. Since you will have to do this after you have paid for the cows, you do run the risk of loss if any are condemned.

SEASONALITY IN FEEDER CATTLE PRICES

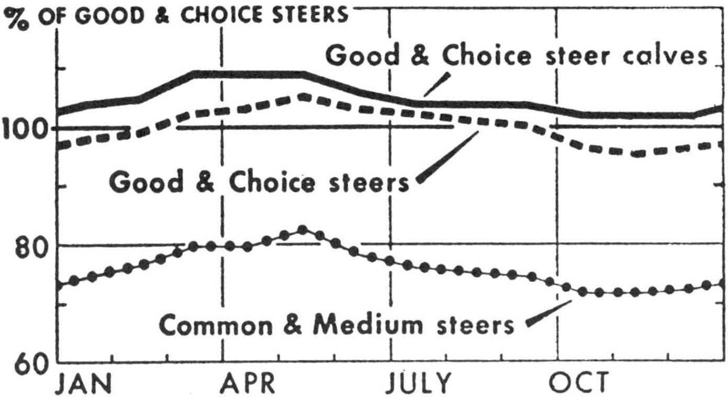

Fig. 1.15—Note that feeder cattle prices are usually lowest in the fall and highest in the spring because of supplies on the markets.

Buy grade cattle at home—Often you can find good bred beef cows and open heifers in some localities. You won't have to test these cattle because they probably have had a county tuberculosis and brucellosis test recently.

Pregnancy Test Replacements—Animals kept as replacements from your own herd and especially animals purchased as bred heifers or cows should be pregnancy tested to determine if they are with calf. It is uneconomical to keep animals to the calving period and not have them produce a calf. If they are not pregnant they may be difficult breeders anyhow and should be culled. In addition, good feed and market conditions can be taken advantage of if a cattleman knows in advance of calving date he plans to cull an animal. Veterinarians and qualified technicians can successfully pregnancy test females. A nominal fee is charged, generally around one dollar per head when a number are done at one time.

Courtesy, Bureau of Sport Fisheries and Wildlife
Photograph by Julian A. Howard

Fig. 1.16—A longhorn steer. This is a classic example of an animal adapted to his environment. In the early West, the longhorn breed was able to defend himself against beasts of prey and a rigorous climate without shelter. The steer above was 17 years old when photographed and brought the highest price ever paid for a longhorn—$800.00. Longhorns today bring prices competitive with other breeds; and in recent years they have recaptured much of their popularity when used in crossbreeding.

Buy Animals Adapted to Your Climate—Animals that come from a climate similar to the one in which they are going will have less stress put upon them so they will ship well and have little setback. Furthermore they probably will be immune to most diseases common to your area because they have already been exposed to them since calfhood and, if need be, already vaccinated.

Select Most Popular Breed in Your Area—There is a tendency to start with strange breeds thinking a monopoly would be advantageous. In most cases quite the contrary outcome results. While a monopoly exists there is often no demand for stock, so marketing becomes a problem. If a well-known popular breed is selected, excess animals are easy to sell and a larger selection is available when buying.

Research Is Needed

While great progress has been made scientifically and many producers are operating at peak efficiency levels, improvement in the beef industry is still needed.

All applied research in the area of livestock production and marketing is important; however, the following list indicates subjects on which emphasis is needed.

1. *Studies on reproduction.* Of particular interest are testing materials or methods including general management for bringing about consistent conceptions, getting cows to breed back quickly after parturition, and encouraging production of *twins* by either selection or hormonal procedures.
2. *Stress investigations.* Needed here are management methods to expedite recovery of severely stressed animals (particularly young calves) after weaning and shipment; use of treatment schedules involving various drugs to promote faster recovery of sick animals and the use of dietary regimes leading to return of the stressed animal to normalcy and promoting its ability to consume feed and gain weight in the usual manner. For example, the *Rumen count* (bacteria) is down after shipping.
3. *Performance of "exotic" cattle crosses* of purebreds as compared with older resident breeds and crosses. This involves performance on the range and in the feedlot, and the resulting carcass.

4. *Early weaning and intensive care operations* with calves or lambs. Such studies are paricularly important if the industry is to make greater use of drylot facilities and more intensive operations.
5. *New approaches to manure management* and new or better developed disposal outlets to minimize the problem of animal waste.

Chapter II

SELECTING THE BREEDING STOCK

Never before in the history of the United States has there been such a variety of cattle breeds available for use in beef production. In order to provide factual information on the performance of these breeds, three researchers from the University of California—Nelson J. Adams, W. N. Garrett, and J. T. Elings—carried out an investigation and reported the results at the California Feeders' Day.

In addition to the standard breeds of British origin which make up the majority of our beef cattle population, there are the Brahman, the Santa Gertrudis, the dairy breeds, and the latest on the American scene—the recently imported breeds termed "the exotics."

Publicity and promotion of the exotics have dominated the trade media, yet information about their ability to perform and to yield desirable carcasses under our range and feedlot conditions is limited. Therefore, an early evaluation of the performance and carcass characteristics of these breeds is essential if they are to be used to maximum advantage by the beef cattle industry.

An evaluation was made of the performance and carcass characteristics of a sample of steers produced by

Fig. 2.0—Limousin cow and calf. This is a French breed of beef cattle, one of the growing number of new breeds imported into the United States. In general these cattle are large and heavily muscled, although often not so smooth as English breeds.

Hereford dams and sired by bulls of the Simmental, Limousin, Maine-Anjou, Lincoln Red, Brown Swiss, Charolais, and Angus breeds. All groups gained well and were efficient converters of feed under the conditions of the test. The Simmental and Maine-Anjou excelled in gain, but no group showed any particular advantage in feed conversion.

Carcass characteristics of the large breeds, Charolais, Simmental, Main-Anjou, and the Limousin, were similar, with these breed crosses yielding a high percentage of edible meat, little fat trim, and—with the exception of the Simmental cross—most failed to grade USDA Choice under the conditions of this test.

The Brown Swiss crosses combined fast and efficient gain with Choice carcass quality to provide the highest net return of any of the groups tested.

Based on this sample, these new breed crosses com-

PERFORMANCE FIGURES OF EXOTIC CROSSES

	Angus cross	Charolais cross	Brown Swiss cross	Simmental cross	Limousin cross	Maine-Anjou cross	Lincoln Red cross
No. head	10	10	9	10	10	10	9
Conformation score[a]							
Beginning	87.7	87.2	87.5	88.0	88.2	87.2	87.1
Final	87.3	87.9	88.0	89.6	88.6	88.0	87.5
Live shrunk wt.[b]							
Beginning	568	622	588	644	563	534	516
Final	956	1011	1040	1135	988	1025	995
132-day performance							
Gain	388	389	452	491	425	491	479
Av. daily gain	2.94	2.95	3.42	3.72	3.22	3.72	3.63
Feed conversion[c]	7.54	7.28	6.91	6.90	6.94	6.16	6.82
Final data							
Live shrunk wt.	972	1026	1076	1154	1008	1059	1021
Total gain	404	405	488	510	445	524	505
Av. days on feed	145	153	160	162	162	162	153
Av. daily gain	2.79	2.65	3.05	3.15	2.75	3.23	3.30
Feed conversion[c]	7.81	8.06	7.65	7.89	7.90	6.92	7.50

a Conformation score: 86-88 = Choice; 89-91 = Choice+.
b Weight taken after 12-hour stand off feed and water.
c Feed conversion, pounds of feed per pound of gain.

Mean values for 132-day gain, final gain, and feed conversion

California Feeders' Day, University of California.

Fig. 2.1—A young cow shown after calving. This is a first cross Simmental Hereford animal exhibiting characteristics of each breed and possessing a high degree of hybrid vigor. A third breed may be used on crossbreeds like the above cow for the next calf crop.

pare favorably with breed crosses presently being used. In the feeding and carcass phase of the industry, it appears they can make a contribution. One of the favorable assets of these new breeds may be that they offer the producer a wider choice of genetic material with which to capitalize on hybrid vigor.

Desirable Breeding Stock Essential—The aim of every breeder is to produce cattle capable of making the highest quality of beef with the most economical use of feed. In order to accomplish this, the selection of desirable breeding stock is a factor of prime importance. Many beef producers are unsuccessful because they did not select the right kind of breeding stock to form their foundation herds. Young breeders and those just entering the field, oftentimes, in their enthusiasm to start operations, pay too little attention to the kinds of animals that will best suit their condi-

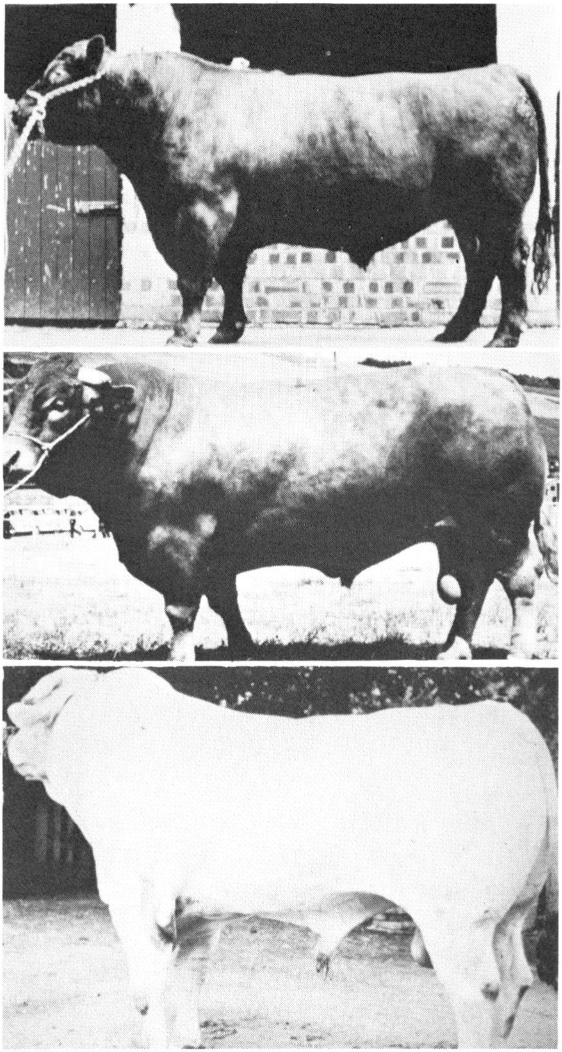

Fig. 2.2—Three breeds recently imported to America, termed "exotics." (Top) Lincoln Red bull from England; (center) Gelbvieh bull from Germany; (bottom) Chianina from Italy. These breeds may have a special influence in crossbreeding programs.

tions. Breeding stock should be so selected to produce a high percentage calf crop of thrifty, healthy, early maturing quality beef animals. Beef producers will be more successful if they recognize and follow approved practices which have been thoroughly tested by experiment stations and farmers and have proved to be desirable for use in the local community.

Activities Which Involve Approved Practices

1. Selecting an approved type.
2. Selecting a suitable breed.
3. Selecting desirable foundation cows.
4. Selecting a desirable herd sire.

1. Selecting an Approved Type

The beef animal should appear to be massive and blocky from all points of view. The body should be wide, deep, yet compact and balanced. Flesh should be firm and thick. The animal should stand squarely on rather short legs and be good headed. It must be able to utilize feed efficiently, produce offspring each year, and have an ample supply of milk to raise a healthy calf.

Two Types of Beef—There are two types of cattle used for producing meat known as "dual-purpose" and "beef-type." By far the most important are the beef type. Some of the common beef breeds are Aberdeen-Angus, Hereford, Polled Hereford, Shorthorn, and Polled Shorthorn. The "dual-purpose" breeds are Milking Shorthorn, and Red Polled. Dual-purpose are used where both milk and beef are to be marketed. They are few in number and seldom used for beef production entirely.

Two Types of Production—In greatest numbers are

the producers who carry on breeding operations to supply the beef market. These cattle are for the most part unregistered. Breeders select foundation females from high quality commercial cows. Established breeders select out of their own herds, for replacements. New breeders generally purchase yearlings or bred heifers. High quality registered bulls are used for sires.

The second class of breeders supply the first group with purebred bulls. These breeders carefully mate their animals, all of which are registered, in a constant effort to improve the breed. New stock is introduced generally by purchasing heifer calves and eliminating unworthy individuals, or by obtaining yearlings or heifers old enough to breed. For the beginning breeder,

Courtesy, Daniel Cassard

Fig. 2-3—Here is a group of excellent purebred Brown Swiss two-year-old heifers living at 12,000 feet altitude on the southern Sierra in Peru. They are part of the herd of the university at Chuquibambilla. Brown Swiss tend to be a dual purpose breed, but here they are used solely for beef and are a good example of cattle adapting to their environment. This breed seems to be one of the few breeds that do well at extreme altitudes.

the purchase of bred cows with calves at foot is a good practice.

Select Adapted Type—Because consumer demands have changed, the market types of cattle have changed. The types of cattle selected for breeding should be early maturing and rugged cattle that will weigh between 700-1000 pounds at 12 to 18 months under good conditions. Rugged, healthy, vigorous cattle are always to be selected regardless of breed. Breeding cows should be selected on their ability to produce a calf each year and give plenty of milk to feed it. It is most desirable to select a type of cattle that have the ability to "hustle" their own feed and do not have to be hand fed.

Some cattle are able to withstand heat and pests

Fig. 2.4—This Brahman bull represents an excellent example of a type of cattle adapted to specialized conditions. Even in humid, hot climates with danger of parasite infestation these animals thrive.

better than others; therefore the most adapted kind of cattle should be selected depending on the graphic location of the ranch. There are differences among breeds that make some breeds more adaptable to a given area than others.

2. Selecting a Suitable Breed

Choose Adapted Breed—The predominating beef breeds of cattle in the United States are Hereford, Shorthorn, and Aberdeen-Angus. In recent years various crosses of Brahman cattle have become important especially in the south. The first two breeds include both horned and polled strains, whereas the last breed is strictly polled. Another recent development

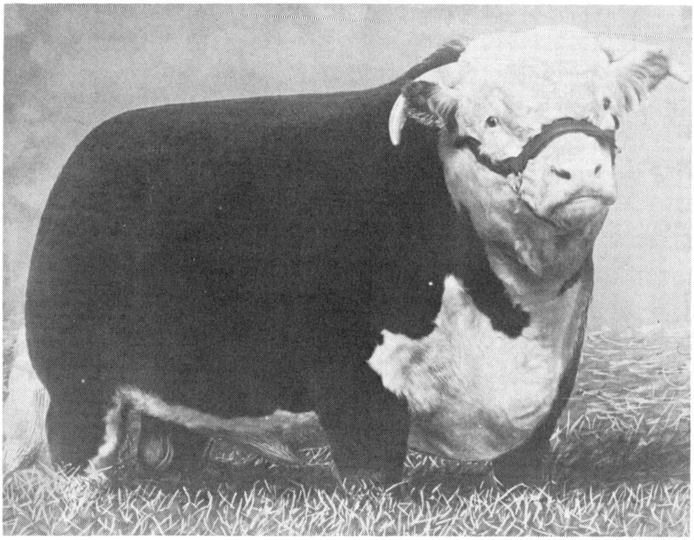

Courtesy, American Hereford Association

Fig. 2.5—An excellent representative of the Hereford breed of cattle. The above animal was champion bull at the World Conference Hereford Show in Des Moines, Iowa.

has been the increased popularity of polled cattle particularly with Polled Herefords. The breed to select is largely one of personal preference. It is generally recognized that there is no one best breed of cattle. There is, however, considerable difference among individuals and strains within a breed. In some communities one breed may be more popular than other breeds. This may mean that this breed is best adapted to local conditions. If this breed meets with the preference of the grower, it will probably be desirable to select it. It is then possible to exchange breeding stock easily and to have cooperation among the breeders in the community.

Herefords

It all started in Herefordshire, England, with a stalwart son of a cow named Silver. Silver was given to Benjamin Tomkins at his father's death. Several years later Benjamin took a bull calf from Silver and bred him to two cows, Pigeon and Mottle. From this union the Hereford breed of cattle emerged.

Also following the same idea was William Gallier who worked with Benjamin in his Hereford breeding program. It was a sale by Gallier in 1795 which was the first fully recorded Hereford auction.

A product of necessity, the Whitefaces were bred to meet the expanding need for beef in England during the 18th century. Designed primarily and exclusively for the production of beef, the Hereford eventually attracted attention in the United States where cattlemen were seeking a beef animal which would rustle for itself on America's plains, hills and mountains.

Henry Clay imported the first Herefords to this nation in 1817. In 1840 the first breeding herd of Whitefaces was established near Albany, New York. Prize winning Herefords were found at eastern county fairs by 1847. The East took Herefords into their herds, their reputation spread westward, and the cattle soon followed by rail.

A registry and promotion organization was founded in 1882 named American Hereford Association.

In 1920, total registrations reached 125,000. By 1930, Hereford breeders had recorded 1,880,000 Whitefaces. Ten years later, the total soared to 3,130,000. When the books were closed in 1951, total registrations passed the seven million mark. By 1964 Hereford registrations had neared the 13 million mark.

Hereford is strictly a beef breed and has met with unusual favor as a rustler in the range areas of the west, northwest, and southwest where it exceeds in numbers all other beef breeds combined.

Polled Herefords

In 1901 the first Polled Herefords were registered—11 head—and from this small start the American Polled Hereford Association has recorded some four and one-half million as of September 1979.

Five Midwest horned Hereford breeders noted with keen interest about 1899 that from registered horned Hereford herds occasionally a hornless calf (with horned parents) was born. This change from parents' characteristics is known as a "sport" or "mutation."

These pioneer breeders, headed by Warren Gammon, of Des Moines, Ia., saw the advantages of developing a Polled Hereford breed after finding the polled, or naturally hornless characteristic, is dominant over the horned trait in breeding.

Mr. Gammon located 13 head of these "mutations" or so-called "freaks of nature" in 1901 and bought them. They were registered with the American Hereford Association, but because they were not identified

Courtesy, American Polled Hereford Assoc.

Fig. 2.6—Polled Hereford female. Calves from this breed are always hornless.

as to their polled character, the American Polled Hereford Cattle Club (forerunner of today's American Polled Hereford Association) was organized for this purpose, to register and promote Polled Herefords exclusively.

Shorthorns

Shorthorn cattle originated in England in the counties of Northumberland, Durham, York and Lincoln several centuries prior to their final development in or about the Teeswater Valley. It is claimed that our Shorthorn type of cattle existed on the Yorkshire estate of the Earls and Dukes of Northumberland for a period of two hundred years prior to 1780. It was not

Courtesy, American Shorthorn Breeders' Assoc.

Fig. 2.7—An outstanding Polled Shorthorn and her first calf.

until about 1700 that the breed began to take its form as an individual breed of cattle.

From this beginning evolved the present day Shorthorn. Mating the best to the best and saving the best for further improvement was the theory of these early improvers: Bakewell, Colling Brothers, Bates and Booth.

The thought of registration came at about this time. The Coat's Herd Book of Great Britain was founded in 1822, and the American Herd Book in 1846.

Shorthorns were the first of the pure breeds to come to America. The first importation landed in Virginia in 1783. Shorthorns soon spread over the entire United States and were popular with the early settlers, providing beef, milk, leather and power for the wagon and plow. The first polled mutants of the Shorthorn breed were the twin heifer calves, Millie and Nellie Gwynne, calved in 1881, whose sire and dam were both registered Shorthorns.

All the beef breeds and most of the dairy breeds have their origin in the British Isles. The only creative feature that can be claimed by American breeders is the polled head. Foreign demand for polled heads is strong, especially in Australia.

Shorthorn color is anything between pure red and pure white. Both the red and white colors will breed true. Roan, a much desired color is simply a mixture of red and white hair and cannot be made to breed true. When two roans are mated, the offspring will average fifty per cent roan, twenty-five per cent red and twenty-five per cent white. Roan cannot be found in any other breed. The predominant choice of color is red and dark roan.

Beef Shorthorns are well distributed throughout the United States, the greatest number being found in the corn belt area. Shorthorn cows are usually good milkers and accordingly produce rapid growing calves. In size they are not surpassed by any beef breed.

Angus

The early history of Aberdeen-Angus, like that of many of our domestic breeds of cattle, is lost in the unwritten annals of the agricultural era of the 17th and 18th centuries. It appears, however, that the Angus breed sprang from the wild native hornless cattle of northern Scotland. The best available evidence indicates that the fountain heads of the breed were in the Brechin district of Forfarshire and the Buchan district of Aberdeenshire.

In these two adjacent counties of Aberdeenshire and Angusshire, formerly called Forfarshire, two local

Fig. 2.8—Excellent type of Angus steer.

bovine strains were developed, known in Buchan of Aberdeenshire as hummel or humlies (hornless), and in Angusshire as dodded or doddies. These districts lay along the route of the old-time cattle trails between the northern grazing grounds and the noted southern finishing grounds of the English markets. From the polled cattle of Angus and the hummel cattle of Aberdeen there can clearly be traced the present Aberdeen-Angus breed and it is from these cradles that the breed takes its name.

We find the representatives of Angus and Aberdeens exchanging first place in equal regularity at the various shows. At Perth, 1829, the breed was designated Angus; at Inverness, 1831, Aberdeen; at Aberdeen, 1834, Aberdeen and Angus, some specimens being actually from Forfarshire. The actual welding of the breeds, however, took place when specimens of both were mated together knowingly—as at Tillyfour by

Mr. McCombie—and at a much earlier date at Ardovie by Mr. Fullerton, owner of Black Meg, which animal is termed by historians the foundress of the breed. Mr. Fullerton in a communication in 1879 said: "Give me 10 good old-fashioned cows of the Buchan kind and let me put an Angus bull with the best head and neck which can be found to them, and I'll venture the off-spring will all take prizes."

In 1873 there could not have been found in the United States, as far as can be traced, a single representative of the Aberdeen-Angus breed, but by the mid-20th century, there had been placed on the roll of the American Aberdeen-Angus Herd Book more than 2,000,000 purebred animals. To this must be added the unnumbered thousands of non-registered representatives of the breed.

The first importation of Aberdeen-Angus into the United States was made in 1873. In that year George Grant, a native of Banffshire, Scotland, and a retired London silk merchant then living in Victoria, Kansas, imported four bulls, two of which he exhibited at the Kansas City Fair that same year, the first polls to appear in an American show yard.

Angus are distinguished by their black color, comparatively smooth coats of hair, and polled character. These cattle are good foragers and able to adapt themselves to varied conditions in the United States. They cross well with other breeds and hold an enviable record in the feed lot and as fat slaughter cattle.

Brahman

The cattle of India known as *Brahman* in the United State and referred to as *Zebu* in Europe and South America are the oldest existing domestic cattle of the world. They have been bred for many thousands of years in their native land of India and are the most numerous of any type of cattle . . . more than one-half of all the cattle of the world possess some Indian *Bos Indicus* blood.

The first Indian cattle of which there is any record in the United States were imported in 1849 by a South Carolina planter named James Bolton Davis. One version is that these animals came from a London Zoo and that the herd consisted of two bulls of the Mysore breed.

In 1854, Richard Peters of Atlanta bought the herd and crossed them with native cattle. It is believed that all purebreds of this first importation were wiped out during the Civil War and so would have made little direct contribution to the present-day Brahman breed.

Next known Brahmans to reach this country were two bulls given to Richard Barrow of Baton Rouge, Louisiana. This was in the year 1854, a gift from the British crown.

The next introduction of imported Indian blood was made in about 1895 when a cow which had been imported by Haggenbach, and which was being shown at Columbus, Texas, gave birth to a bull calf that was sold to a Mr. Hahn near Houston.

Thousands of cattle of India were imported into Brazil, where they are known as Zebus and thrived under Brazilian conditions. The popularity of Brahmans in Texas led Dr. F. Ruffier, acting as agent for members of the Sociedade de Herd Book-Zebu, Uberaba, Brazil, to export from Brazil to Texas by way of Mexico, about 90 bulls of the Guzerat, Gir and Nellore races. John T. Martin of San Antonio assisted in financing the bringing of these bulls to Texas. They were generally distributed among the breeders of Brahman cattle of South Texas. This importation reached Texas

Fig. 2.9—Brahman cow. This breed originally came from India, where it is known as Zebu. It has been imported into all countries of the world and has made a great impact on size and appearance of other breeds, especially in warm climates.

in 1924, the year the American Brahman Breeders Association was founded in Houston, Texas.

From these early beginnings, the number of Brahman cattle in the United States has grown with astounding rapidity.

Brahman cattle are characterized by a prominent hump above the shoulders, abundance of loose pendulous skin under the throat, on the dewlap, navel, and sheath. This large amount of loose skin gives a greater surface area to the animal; therefore assists in cooling. The primary purpose in introducing Brahman (native of India) cattle into the United States was to cross on other breeds in order to develop a type that would be more adaptable to the Gulf Coast region. These cattle seem to endure hot weather very well even without shade, and are able to travel long distances to water.

Galloways

Although the breed's origin is shrouded in antiquity, there is good reason to believe Galloways stem from the species *Bos Longifrons* that roamed wild in the dense forests of Scotland prior to the Christian era. Among those ancient cattle were the Celtic dark-colored cattle, domesticated in Britain even before Caesar's invasion

Fig. 2.10—Galloway cow. A common sight in Scotland but at present not as well known in America.

in 55 B. C. It is reasonable to believe that is the source of Galloway color.

The breed, as the name implies, is native to Galloway, that southernmost extremity of Scotland's Lowlands which juts down to Mull of Galloway and Barrow Head.

Canada's First Importation. The first Galloways to cross the Atlantic were brought into Canada by a Mr. Graham of Ontario in 1853. The shipment comprised a yearling bull, Jock 10—the first bull recorded in Vol. 1 of the North American Herd Book—and nine females, numbered 1 to 9 inclusive in that volume.

America's First Galloways. The first Galloways ever to tread American soil were brought in from Ontario in 1866 by Michigan State College, Lansing. The shipment comprised a two-year-old bull, Victor 122, and a heifer of approximately the same age, Blooming Heather 80.

America's First Importation. Peter Davy, Dodge County, Wisconsin, was breeding Galloways as early as 1874 and was one of the most extensive recorders of Galloways in Volumes I and II of the American Galloway Herd Book. In the early 1880's Mr. Davy made the first importation directly from Britain to America. The shipment consisted of three bulls and 16 females purchased from Scotland's leading breeders.

American Galloway Herd Book Born. On November 24, 1882, during the Fat Stock Show held in Chicago, a meeting of Canadian and American breeders formed an association to include breeders throughout North America. At this meeting it was decided to publish a herd book. Thus, Vol. I of the North American Herd Book was born. Later, when the American Galloway Breeders' Association was formed, that book was adopted as Vol. I of the American Galloway Herd Book.

In 1911, Vol. XVIII was published recording about 35,000 animals. Lately, activity is increasing as the numbers grow and attract new breeders.

Belted Galloway

The recognized colors are black with brownish tinge and a white belt completely encircling the body between the shoulders and the hooks, or dun with a similar belt.

Origin of the Belt. Seeking in our agriculture literature for the origin of this striking coloration, it appears that "belted" or "sheeted" cattle were found at various times not only in Galloway and northern England, but also in Aberdeenshire; East Anglia; Yorkshire; Worcestershire; Somerset; Wales and Ireland. Some of these cattle were polled and some horned, which seems

Courtesy, Aldemere Farm, Rockport, Maine

Fig. 2.11—Belted Galloway. As yet these cattle are not common in America.

to have been the case with Galloway cattle about a hundred and fifty years ago, and it is probable that Belted Galloways like all our so called pure breeds are genetically heterogeneous in origin.

In Holland there is a breed of black belted cattle called Lakenvelders of which two herds still survive. The Dutch do not know the origin of these cattle, which seem unlikely to have been derived from their ordinary white and black cattle. The Belted Galloway might be derived from a cross between imported Lakenvelders and native Galloway cattle, and if this is so, then the Beltie is indebted to Holland for a strain of Dutch blood in the breed in the same way as the Shorthorn, Aberdeen Angus and Hereford are thought to be.

Charolaise

Developed during the 17th century in the hill country of central France, the Charolaise is recognized today as a highly specialized beef breed.

In the early 1900's the first importation of these cattle reached the North American continent and continued until 1937. These first imported cattle were brought into Mexico, and it is from these herds that breeders in the United States secured their foundation stock, beginning in the late 1930's.

The first of the sturdy Charolaise breed of cattle reached the United States through Texas, where pioneers in this breed of cattle began a systematic crossbreeding program that led to the establishment of the Charbray breed—a cross between the Charolaise and the Brahman.

The Charolaise (other spellings Charollaise, Charollais) is one of the most important breeds of French cattle. It is white or light creamy colored and origi-

Fig. 2.12—Charolaise beef cattle have only reached the United States since the 1900's.

nated in the Province of Charollais. It is a popular breed in the prairie section of central France. It is one of the largest of all breeds of beef cattle.

Devon

The color of Devon cattle is a dark cherry-red sometimes referred to as "Red Rubies." They are horned and have a reputation for being good milkers but are not a dual-purpose breed.

Little is known of the exact origin of the breed. Charles Kingsley introduced them into a scene in his famous novel "Westward Ho!" and there is a record of some red cattle being taken from Cornwall to Kerry in 1580. Colonel J. Tanner Davy, of Rose Ash, near

Fig. 2.13—Devon bull. Devons are large, dark red cattle and are extensively used in hybrid or crossbreeding. In England frequently two calves are raised on one Devon cow. They are an important member of the new exotic breeds to invade the U.S.

South Molton, Devon, who founded "Davy's Devon Herd Book"—the official record of the breed—stated in 1851 that his family had been breeding choice Devons for one hundred and fifty years, while members of another pioneer breeding family, the Quartlys, are known to have had a herd of "Red Rubies" at Molland, North Devon, in 1703. Present-day catalogues include animals which trace their ancestry to early "Davy" and "Quartly" strains.

Red cattle of Devon breeding were brought to North America in 1623 by the Plymouth Colony.

Although not as numerous as other breeds their distribution is world-wide. In the U. S. most specimens are found in the south especially in Mississippi.

In 1905, when the American Devon Cattle Breeding Association was formed, the American Devon Record became the official register. The association is now organized as the American Devon Cattle Club, Inc.

Simmental

The Simmental is one of the oldest and most widely distributed of all breeds of cattle. Originating in the Simme Valley in western Switzerland, the breed, under a variety of local names, has spread throughout Europe and has played an important role even in countries of the mid-East, in Russia, in South Africa, and recently has been introduced into South and North America.

Long before the advent of any herd book, evidence of the productivity of these red and white cattle appeared in ecclesiastical and secular property records in the western Swiss cantons. As early as the 18th century, before the idea of district breeds was conceived, they were in great demand in Lombardy where their propensity for rapid growth and good finish were plainly recognized. They were also much sought after because of their outstanding production of milk, butter, and cheese and their utility as draft animals. Year after year emissaries of central and eastern European princes came to the valleys of western Switzerland, to the Simme, the Sarine, the Kander, and the upper Aar, to purchase bulls, dairy cows, beef cattle, and draft animals which they selected from this strain.

Eventually the breed was formally established. In 1806, the Bernese government (the government of the canton, or state, of Berne) set up a herd register. And with the register came the requirement of a performance pedigree, one of the first in connection with any herd book. From that time on it has been required that all Simmental cattle of breeding age must be appraised by an official commission of judges before entry in the herd book. Only when their pedigree, their performance, and their conformation meet the required standards are they accepted.

On October 14, 1968, the American Simmental Association was formed. Its primary objective is the development, registration, and promotion of the Simmental breed in the United States. The American herd book is based on performance pedigrees, and in all breed promotion emphasis is placed on performance for production and profitability.

Country:	Local Name:
Switzerland	Simmental
Austria	Austrovieh
France	Pie Rouge
Germany	Fleckvieh
Italy	Pezzata Rossa Friulana

Scotch Highland*

The Scotch Highland breed of cattle, developed in the Hebrides, is reputed to have been recorded as a definite breed in the 12th century. In size or body weight, it is considered one of the smaller or, at most, intermediate breeds. Owing to continued selection under adverse conditions involving rigorous climate and at times scant feed supplies, animals of this breed have inherent hardiness coupled with mothering ability. Acceptable colors are black, brindle, red, light red, yellow, dun, and silver. The horns are wide and branching. The ears are quite often short. The hair coat has been appropriately described as a "coat and vest," the undercoat being soft and thick while the outer coat is coarse and long. They have straight lines and in their original home they are reputed to produce excellent carcasses of high quality.

The American Scotch Highland Breeders Association was formed in 1948.

Sussex*

The Sussex breed originated in the counties of Sussex and Kent in southern England. It is considered to be one of the oldest breeds developed in Great Britain. Originally, Sussex cattle were used for draft purposes, which undoubtedly accounts for their relatively large size and good feet.

The Sussex is noted for its high average daily gain. Typical representatives are horned, and dark red in color.

Outside of its country of origin, the Sussex has at-

Fig. 2.14—Scotch Highland steer. This animal is three years old with a shaggy long coat of reddish brown hair and a huge set of horns. There are few imports into the U.S.

*USDA Bulletin 1779

tained its greatest prominence in the Union of South Africa, in southwest Africa, and in Rhodesia. It has crossed well with the native Africander. Only a small number of Sussex have been introduced into the United States.

Courtesy, N. W. Higgins, Somerset, Calif.

Fig. 2.15—A Beefmaster cow and calf. This is a relatively new breed of cattle, reddish brown in color and quite large and rugged. The Brahman and Shorthorn background is evident in drooping ears and rugged frame.

Breeds Developed in the United States

Beefmaster*

In 1908, the late Edward C. Lasater of Falfurrias, Tex., who was the ninth president of the Texas and Southwestern Cattle Raisers' Association, began using Brahman sires on a large herd of Hereford and Shorthorn cattle. He used Gir, Nellore, and Guzerat Brahmans. His Herefords were selected for red pigmentation around the eyes and for milk production.

After Mr. Lasater's death on March 20, 1930, his son, Tom Lasater, began to combine the entire breeding herd into what is now known as the "Lasater Beefmaster," a trade-mark which has been copyrighted through the United States Patent Office. It is estimated that the Beefmaster contains half Brahman, one-quarter Hereford, and one-quarter Shorthorn.

While no specific color requirements have been established in the selection program, the majority of Beefmasters are dun, brown, reddish-brown, or red with some white extensions and spots in certain individuals.

As is to be expected from their genetic background, they are relatively large cattle.

Brangus

Many exciting and important things have happened in the cattle business in recent years as a result of continued experimenting in an attempt to develop cattle that will more economically produce better beef. One of the most significant of these developments has been the Brangus breed, which is fast becoming one of the most popular in the country. The romantic story of the development of Brangus cattle is one which is of interest to all, and is the story of Frank Buttram of

*USDA Bulletin 1779

Fig. 2.16—Brangus cow and her calf. Both the Angus and Brahman were instrumental in the development of Brangus cattle. Before an animal can be recorded by the International Brangus Breeders Association (IBBA) it must pass a visual inspection.

Oklahoma City, Oklahoma. Other cattlemen such as Tom Slick of San Antonio, Texas, also figure prominently in the development of the breed which came into being around 1940.

Brangus is a combination of the Brahman and the Aberdeen-Angus. To be more exact, it is three-eighths Brahman and five-eighths Angus. To begin with, a purebred Angus is bred to a purebred Brahman, producing a half-blood Brahman-Angus. This half-blood is bred back to a purebred Angus, giving a quarter-blood, which is one-fourth Brahman and three-fourths Angus. These two are then bred together to give the Brangus. Later developments have shown that it makes no difference whether an Angus cow is bred to a Brahman bull, or a Brahman cow to an Angus bull; their calves are virtually identical.

Another route which is sometimes called the "short-cut" route is to take a half-blood cross and breed to a Brahman, giving an animal that is three-fourths Brahman and one-fourth Angus. These three-fourths bulls can then be used with Angus herds, giving Brangus

ORIGINAL ROUTE TO PRODUCE BRANGUS

Purebred Brahman
X
Purebred Angus
→ ½ Blood
(½ Brahman — ½ Angus)

½ Blood
(½ Brahman — ½ Angus)
X
Purebred Angus
→ ¼ Blood
(¼ Brahman — ¾ Angus)

½ Blood
X
¼ Blood
→ Brangus
(⅜ Brahman — ⅝ Angus)

SHORT ROUTE TO PRODUCE BRANGUS

Purebred Brahman
X
½ Blood
(½ Brahman — ½ Angus)
→ ¾ Blood
(¾ Brahman — ¼ Angus)

¾ Blood
X
Purebred Angus
→ Brangus
(⅜ Brahman — ⅝ Angus)

Fig. 2.17—Either of above steps can be used to develop your Brangus herd. Thereafter, Brangus bred to Brangus produce Brangus.

calves directly from Angus cows. The name is a little misleading in that it is not a shortcut because it still requires three crosses to arrive at Brangus. However, one of these steps is taken by the breeder of the bull and the purchaser of the bull is saved this step, thus the origin of the term "shortcut." It should be pointed out that an animal with some Brahman and some Angus blood is not a Brangus. The proportions must be three-eighths and five-eighths and no other. One of the biggest problems of the Association today is that any Brahman-Angus cross is called a Brangus.

Charbray

This comparatively new development in the beef cattle industry came about as the result of crossbreeding of the Charolaise, the long-established beef breed

Fig. 2.18—Charbray cattle are a new development in the beef cattle industry. The crossbreeding program in Texas led to the development of the breed.

of France, and the Brahman, for years a popular animal in beef herd crossbreeding programs in this country.

The crossbreeding that led to the Charbray cattle was started in the Lower Rio Grande Valley in Texas in 1944. Pioneers in the program that led to the establishment of this breed used Charolaise cattle registered in the American Charolaise Breeders Association and registered Brahmans.

Crosses with other beef breeds were carried out, and though they showed remarkable calves, the Charolaise-Brahman cross proved to bring the sturdiest animals with ability to gain weight faster.

In color markings, Charbray animals resemble the Charolaise in that they are usually a creamy white. They are horned and while they show almost no evidence of the Brahman hump they do show slight evidence of the Brahman dewlap and some excess skin in that region. They are large cattle and produce vigorous, fast-growing calves.

Santa Gertrudis*

The Santa Gertrudis breed of beef cattle was developed on the King Ranch in Southeast Texas from a cross between beef type Shorthorn cows and beef type Brahman bulls. Crossbreeding between these breeds of different species, *Bos taurus* and *Bos indicus,* started in 1910 and continued for approximately 15 years. The breed first started with the bull Monkey. His sons and grandsons were used in the herds as rapidly as possible until eventually only bulls which descended from him were used. After 30 years of following a well-defined plan of selection, inbreeding, and linebreeding of the offspring of Monkey, the characteristics had become sufficiently well fixed that the United States Depart-

*USDA Bulletin 1779

Courtesy, Santa Gertrudis Breeders International, Kingsville, Texas

Fig. 2.19—A Santa Gertrudis bull. At 33 months of age he weighed 2,010 pounds. The Santa Gertrudis is a large, popular breed developed at the King Ranch in southeast Texas.

ment of Agriculture recognized the Santa Gertrudis as an established breed in 1940. It is the first distinctly American breed of cattle to be recognized. Its name is derived from the Santa Gertrudis division of the King Ranch, where the original foundation cattle were maintained and the breed was developed.

The Santa Gertrudis is a large beef animal with mature cows frequently attaining 1,600 pounds and mature bulls 2,000 pounds in weight.

It is solid cherry-red in color with flesh colored mucosae. It is a horned breed, although polled individuals are sometimes seen. The ears are somewhat large and semi-pendulant. Both males and females are loose-hided, showing considerable dewlap and underline skin folds. A short, straight hair is characteristic.

Dual Purpose Breeds

Several breeds of cattle are listed by breed associations as dual purpose breeds. These cattle are primarily beef animals but in addition are good milk producers although inferior to the strictly dairy breeds. As a rule they raise heavy calves and tend to be used on small farms or where seasonal dairying is practical. Milking Shorthorns and Red Polls are the two best known breeds.

Milking Shorthorn

Selective breeding and improvement of cattle which resulted in the Shorthorn breed was first carried on in the River Tees Valley in northeast England which includes the counties of Durham and York, dating back as early as 1730. The first English herd registry book was published in 1822 with data on cattle dating back into the early 1730's, and of course a complete herd registry has been kept on English Shorthorns since the beginning. Much of the historic credit goes to

Fig. 2.20—A Milking Shorthorn cow. This perhaps is the best known of the dual purpose breeds.

Thomas Bates of Kirklevington who developed the great Bates Line of Shorthorn cattle.

Cattle of this type are referred to as Dairy Shorthorns in England, Dual-Purpose Shorthorns in Canada, Milking Shorthorns in the United States. Some countries call them Durhams because their first importations came from the county of Durham, England. Regardless of what they are called, they originate from the English Dairy Shorthorn.

Shorthorn cattle were first established as a registry organization in 1846 in the United States, and up until the early part of the 20th century, were all recorded as Shorthorns. However, some herds were of the beef type while others were of the Milking Shorthorn type. In 1915, the Milking Shorthorn breeders of America were instrumental in establishing a milk recording organization so that they might have definite milk and butterfat production data on their cattle. In 1920, the American Milking Shorthorn Society was established wholly for the purpose of promoting the interest of Milking Shorthorn breeders and eventually identified all the Milking Shorthorn breeders in America and slowly promoted the breed through a magazine and the milk recording department.

In 1948, the American Milking Shorthorn registry was established which registers all horned Milking Shorthorns and polled Milking Shorthorns in the United States. The parent association, the American Shorthorn Breeders Association, records all the horned Beef Shorthorns and polled Beef Shorthorns.

Red Poll

In the east of England, in the counties of Norfolk and Suffolk (settled by the Englens), there were developed, prior to the year 1800, two distinct strains or "stocks" of cattle. They can hardly be referred to as breeds since no record associations or standard descriptions existed.

The Norfolk cattle are described by the historians as being mostly blood red in color with white or mottled faces and some white on the underparts. They were rather small, but had good carcass qualities and were early maturing. They were mostly horned cattle in Norfolk although the sales of six polled herds in 1802 and thirteen in 1804 have been documented.

The Norfolk "home-breds," as they were called, were used mostly to produce beef, but they were likewise used some in the dairies.

In the neighboring county of Suffolk, the cattle were

Fig. 2.21—Red Poll cow. Cows of this breed are large, red or brindle, and good milkers; consequently they are often used under diversified farming conditions in England.

somewhat larger, red, dun, or brindle in color, all polled and fine milkers.

Red Polls came about through the merging of the Norfolk and Suffolk stocks. It is perhaps significant that the first known improver was himself a tenant farmer, John Reeve, on the Holkham Estate in Norfolk. He used Norfolks and Suffolks about 1800 and utilized crossbreeding in developing the breed which is considered a dual purpose breed. The Red Poll Cattle Club of America was organized in 1883, five years before the English society was formalized.

Today Red Polls are raised in 44 states. From the beginning they have been favored where diversified farming is practiced.

3. Selecting Desirable Foundation Cows

Select Females First—Selecting the right females is a very important initial step. In starting a breeding herd, select the females first and then select the bull

Fig. 2.22—A herd of Murray Grey cattle. The breed originated in Australia.

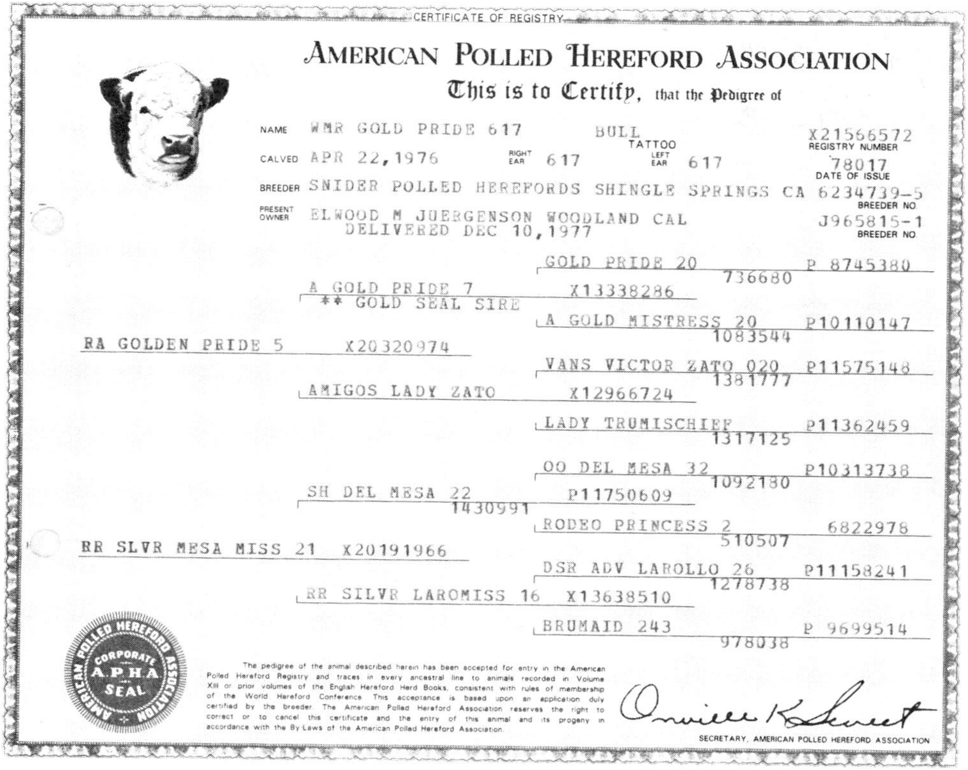

CERTIFICATE OF REGISTRY

AMERICAN POLLED HEREFORD ASSOCIATION

This is to Certify, that the Pedigree of

NAME WMR GOLD PRIDE 617 BULL X21566572 REGISTRY NUMBER

CALVED APR 22,1976 TATTOO RIGHT EAR 617 LEFT EAR 617 78017 DATE OF ISSUE

BREEDER SNIDER POLLED HEREFORDS SHINGLE SPRINGS CA 6234739-5 BREEDER NO.

PRESENT OWNER ELWOOD M JUERGENSON WOODLAND CAL J965815-1 BREEDER NO.
DELIVERED DEC 10,1977

GOLD PRIDE 20 P 8745380 736680

A GOLD PRIDE 7 X13338286
** GOLD SEAL SIRE

A GOLD MISTRESS 20 P10110147 1083544

RA GOLDEN PRIDE 5 X20320974

VANS VICTOR ZATO 020 P11575148 1381777

AMIGOS LADY ZATO X12966724

LADY TRUMISCHIEF P11362459 1317125

OO DEL MESA 32 P10313738 1092180

SH DEL MESA 22 P11750609 1430991

RODEO PRINCESS 2 6822978 510507

RR SLVR MESA MISS 21 X20191966

DSR ADV LAROLLO 26 P11158241 1278738

RR SILVR LAROMISS 16 X13638510

BRUMAID 243 P 9699514 978038

AMERICAN POLLED HEREFORD ASSOCIATION CORPORATE A.P.H.A. SEAL

The pedigree of the animal described herein has been accepted for entry in the American Polled Hereford Registry and traces in every ancestral line to animals recorded in Volume XIII or prior volumes of the English Hereford Herd Books, consistent with rules of membership of the World Hereford Conference. This acceptance is based upon an application duly certified by the breeder. The American Polled Hereford Association reserves the right to correct or to cancel this certificate and the entry of this animal and its progeny in accordance with the By-Laws of the American Polled Hereford Association.

Orville K Sweet
SECRETARY, AMERICAN POLLED HEREFORD ASSOCIATION

Fig. 2.23—Pedigree form used by American Polled Hereford Association.

that will improve them. Following are some specific points to consider:

1. Representative of the breed.
2. Type and individuality.
    a. Type—thick and wide.
    b. Thickness of natural fleshing.
    c. Head and sex character.
    d. Construction and ruggedness.
    e. Quality and smoothness.
    f. Size for age.
    g. Substance and legs.
    h. Balance and symmetry.
3. Breeding record: Calves coming from cows and bulls which regularly produced outstanding calves are to be preferred. Select heifers from cows who have produced a calf each year and have ample milk to insure large calves for a given age.
4. Age to buy: There are a number of successful methods depending on individual situations.
    a. Buy a large number of calves at weaning time and later select.
    b. Buy yearlings or two-year-olds near breeding age.
    c. Buy bred heifers mated to good bulls.
    d. Purchase mature cows with calves at foot and rebred.
5. Pedigree: In selecting for or from a commercial herd, pedigrees are almost always lacking except for verbal information that can be obtained from the breeder. This information is as reliable as the breeder. Generally, more emphasis should be given to the individuality and performance of an animal than to its pedigree. In judging a pedigree, attention should be given to the first three generations. Two types of pedigree forms are used—the extended form used by Aberdeen-

Angus and Hereford, and the abbreviated used by Shorthorn breeders.

From a practical standpoint, the sire is the most important part of the pedigree because more information is available concerning his ability to produce desirable offspring.

6. Health: Always buy healthy cattle. All cattle should pass the T.B. test or come from clean herds. If possible, they should pass a negative test on Bangs' disease or have been vaccinated for the disease as calves.
7. Identification: Check for proper identification whether it be branding, tattoo, or ear notch.

Never sacrifice quality for numbers, as it is always better to buy fewer head of better quality cattle. In the future, "efficiency of gain" will be one of the most important factors in selection of beef cattle.

SULTAN PRINCE 2D 1914193

Roan, calved July 10, 19--. Bred and owned by --------------------
(Name and address)

Dams	*Breeders of dams got by—*	*Sires*	*Breeders of sires*
Edellyn Rosewood 9th 1580535	Thos. E. Wilson	Sultan Browndale 1771330	Thos. E. Wilson
Winmoor Rosewood 4th	W. H. Schendorf	Browndale Count 1156438	James Douglas
Ravendale Rosewood	John O. Pew & Son	Linwood Rodney 1020043	Carpenter & Ross
Rosewood 93d	Carpenter & Ross	Glaryford Augustus 716421	M. Dysart
Maxwalton Rosewood 6th	Carpenter & Ross	Revolution 388359	Carpenter & Ross
Rosewood 92d	W. D. Morrison	Maxwalton Renown 367543	Carpenter & Ross
Rosewood 76th	J. Bruce	Golden Prince 306267	R. Copland
Rosewood 54th	J. Bruce	Waverley 136405	A. Strachan
Rosewood 23d	J. Bruce	Cap-a-Pie 106717	J. Bruce
Rosewood 13th	J. Bruce	Duke of Edinburgh 130843	J. Bruce
		Privy Seal 109647	A. Cruickshank

Sultan Browndale, Sire, Glenburn Field Marshal 1709714 by Edellyn Favorite 1430250, out of Clipper Countess 2d 1215702; dam, Rookwood Golden Chain 5th 1712052 by Golden Browndale 1498465, out of Golden Lady 13th 1307124.

Browndale Count, sire, Browndale 334947 by Avondale 245144, out of Maxwalton Mina 2d 86601; dam, Morning Blossom 2d 1023473 by Roan Chief 361556, out of Morning Blossom 711477.

Linwood Rodney, sire, Rodney 753273 by Sanquhar Dreadnought 680399, out of Rosetta 7th 753278; dam, Baroness Undine 794380 by Baron's Pride 306627, out of Nonpareil Undine 794391.

Fig. 2.24—Pedigree form used by Shorthorn Association.

Get Help in Reading Pedigrees—Until a person gains considerable skill in reading pedigrees accurately it is a desirable practice to seek aid. Pedigrees can be tricky, oftentimes having many well-known names scattered throughout and yet meaning little concerning the animal in question. Some families may be dwarf carriers, or superior animals may be named but be so far removed as to have little value.

4. Selecting a Desirable Herd Sire

Select Bulls for High Production—Real progress in cattle breeding is impossible unless bulls which make each succeeding calf crop better than the last are consistently used. The bull should be selected only after

Courtesy, American Polled Hereford Assoc.

Fig. 2.25—Excellent type herd sire. Consider pedigree, the bull as an individual, and the production record when selecting bulls.

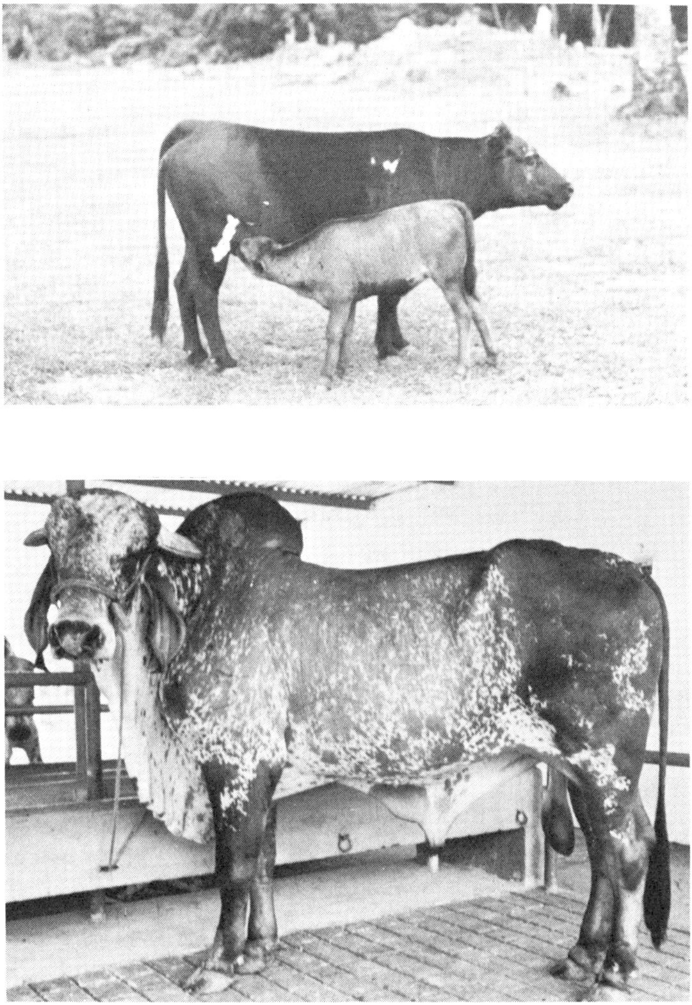

Fig. 2.26—There are many more breeds of cattle in the world than the recently imported "exotics." Above are illustrated two such breeds. Top photo is a Ndama-type cow and calf from Liberia. Below is a Gyr bull from Venezuela. In general, they are mixed types rather than pure breeds, although many characteristics breed true which make them adaptable to their environment.

the cows are carefully studied. Pick a bull whose individuality is strongest where the cows are weakest.

For the purebred herd, more attention should be given to popular blood lines. For the range herd, experience has taught cattlemen that bulls raised in the feed lot and not accustomed to rustling will not do so well when turned out on the range. Following are some desirable factors in a herd sire:

1. Representative of an approved type and breed.
2. Strong masculinity.
3. Normal sex organs, producing viable, mobile sperm.
4. A medium-sized bull is generally preferred to an unusually large or under-sized one.
5. Vigorous and healthy.
6. Strong, straight, squarely-placed legs.
7. Possesses excellent beef conformation.
8. Records indicate superior performance in offspring.

Select Older Bulls on Breeding Record—If a sire has once demonstrated his ability to produce, his pedigree is not especially important. Therefore, select older bulls on their breeding record.

In selecting a young bull, see, if possible, both sire and dam. Consider both pedigree and the bull as an individual. If these are both good, there is an excellent chance he will give a desirable performance.

Select Sire Strong Where Your Herd Is Weak—There is a tendency to attempt to correct all faults at once. A bull that could do this would be rare indeed. The most desirable practice is to correct one factor at a time so that over a period of years, using sires strong where your herd is weak, a superior herd of animals will be produced.

For location of breed associations, see Appendix A.

Chapter

III

BREEDING AND IMPROVING BEEF CATTLE

Establish a Desirable Breeding Program—One of the best ways of improving the beef herd is through the use of a desirable breeding program. This means that animals of an approved and adapted type should be selected. It also means that a number of approved practices should be followed in establishing a desirable breeding program which will result in improving the herd.

In order to be successful, it is very important to have a high percentage calf crop. Use of approved practices can increase the percentage of calf crop.

Activities Which Involve Approved Practices

1. Choose a desirable system of breeding.
2. Keep breeding herd the proper size.
3. Select replacements on a performance record.
4. Breed heifers at the right age.
5. Managing for artificial insemination.
6. Maintain a high percentage calf crop.
7. Select the proper breeding season.
8. Use the right number of cows per bull.
9. Care for the herd bull properly.

10. Care for the breeding herd properly all year.
11. Keep a record of performance.
12. Determine the stage of production to enter.
13. Bulls vs. steers (for beef)

1. Choose a Desirable System of Breeding

Select a system of breeding to meet needs. The breeding system to be followed generally falls into one of

Fig. 3.0—Modern beef breeds were evolved by a selecting and breeding program. The animal shown is a Musk ox and is native to the Arctic Circle, where a breeding program has been initiated to domesticate and upgrade these animals. The purpose of the program is to domesticate musk ox through selective breeding and to develop an economy for Arctic peoples based on its underwool "qiviut." Musk ox presently survive in Greenland and Canada. Wild cows give birth every two years. Domesticated cows breed annually. Adults reach 800 to 1,000 pounds and live approximately 25 years. Up to 6 pounds of qiviut fleece, shed in spring, is pulled by hand off the animal. The fleece is soft gray-brown and can be bleached and dyed. It is one of the world's rarest textile fibers.

two kinds: (1) systems used by commercial breeders, and (2) systems used by purebred breeders.

Buy Bulls on a Grading System

In the past a popular way to purchase sires was to buy bulls who were grand champions or rated high in competition. This practice has been of real value in improving livestock. However a modern, more dependable way is to purchase bulls who have been graded using a recognized grading system such as the University of California Grading System. This is a numerical system that indicates the quality of the bull. For example, a number 1 bull would be a top bull, 2 of lower quality, and so on. However, all bulls rating say 2 would tend to be equal. Under such a system commercial herds would use nothing lower than a 2− bull and purebred herds nothing less than a 2+ bull. Ultimately each herd would attempt to build up to where it would begin using the next step higher grade sire.

Another system would be to purchase bulls on their rate of gain or on the rate of gain they are able to produce in their calves. Bulls selected by this latter system are said to be progeny tested.

At the present time progeny tested bulls are scarce; however, this is the goal for which producers should be aiming. Some livestock men may be able to progeny test their own bulls; if they cannot, oftentimes they can cooperate with their Agricultural Experiment Station in determining which bulls are able to transmit desirable beef characteristics in the calves they produce. In any event an attempt should be made to purchase bulls that have been progeny tested.

W.D.A. is an important measurement to use. This is the weight-per-day-of-age, including birth weight, and is calculated from date of birth to weight date. It is the actual lifetime average daily gain and does not

GRADING GUIDE FOR BEEF CATTLE RECORDS OF PERFORMANCE

Grade		Breeding Cattle	Market Cattle	
Designation	Numerical Value		Feeders	Slaughter
1+ 1 1–	98-100 95-97 92-94	The top of the grade represents outstanding animals in strong competition. The middle and lower ends of the grade represent excellent breeding animals from standpoint of type, conformation, quality and character, capable of making a good showing in strong competition.	Strictly fancy or select	Top Prime
2+ 2 2–	89-91 86-88 83-85	Cows in grade 2+, 2, 2– are good enough to retain for breeding test in purebred herds. This is a practical top for commercial cattle. The top of grade 2+ represents the best of range bulls; the lower end of herd bulls. Cattle in this grade are well down the line or out of the money in strong competition.	Choice	Prime to Top Choice
3+ 3 3–	80-82 77-79 74-76	Cows usually should be culled from purebred and some commercial herds. Cows in this grade are usually the low end of commercial cattle. Bulls are rarely capable of making much improvement except on very plain cattle.	Good to Medium	Choice to Good
4+ 4 4–	71-73 68-70 67	Plain, upstanding, thin-fleshed, slow maturing cattle, lacking in quality and character and having serious defects of conformation. Should be culled from commercial herds.	Medium to Common	Commercial to Utility

Courtesy: Agricultural Extension Service, University of California, Davis

This grading guide illustrates the criteria a livestock specialist would use in evaluating an animal in order to give it a numerical rating.

—*Courtesy, Chas. Pfizer & Co., Inc.*

Fig. 3.1—Here with the carcass is where the "payoff" occurs for all the time, effort, and money which have gone into good feeding, breeding, and management practices. Modern beef production demands that carcass quality and yield of retail cuts be considered in a breeding program.

reflect any adjustment factors. The heritability of W.D.A. is 60 per cent and is one of the most highly heritable traits in beef cattle.

Researchers are exploring the relationship between carcasses of the same grade and their yield in lean meat, and therefore retail value.

U.S.D.A. technicians, in their studies of more than 1,000 carcasses, have demonstrated that carcasses of the same grade and weight can vary widely in their yield of lean meat in the preferred cuts—and hence in their retail value.

Grade—The University of California conformation scoring system has been used in the past. In some places this system has been replaced by the Beef Improvement Federation (BIF) scoring system which is being used instead of the UC system. The two grading systems compare in the following manner:

BIF	UC
17	95-100
16	92-94
15	90-91
14	89
13	87-88
12	86
11	85
10	84
9	83
8	82 & below

The difference in value between carcasses, in extreme cases, can be more than $150. A difference of $50 between carcasses is not unusual. The studies have shown that in higher grades each 1 per cent increase in the carcass yield of the preferred retail cuts increases the carcass value by approximately $1.25 per hundredweight.

As to methods of recognizing these differences in yield, the studies demonstrated that these could be predicted quite accurately by using only four factors:

a. Thickness of fat over the rib eye.
b. The size of the rib eye muscle.
c. The amount of kidney and pelvic fat.
d. Carcass weight.

Even more revealing is comparing animals graded as live animals and then comparing them, after slaughter, as to their yield of retail cuts. All too often fat stock that place high at livestock shows have very disappointing carcasses and low-yielding values in retail cuts. In the future, bulls must be selected to produce calves that yield carcasses high in retail value. The alert producer will recognize these factors in his breeding and selecting program if he is to build and maintain a herd that will bring him the greatest profit.

Methods Used by Commercial Breeders

Grading Up—The most common method of improving a herd is to use a purebred bull on grade cows. This is an inexpensive, sound way because grade cows are used. The herd is improved quite rapidly if a bull with good breeding and desirable characteristics is used. Most obvious improvements in the calf crops are observed on the first cross. A superior sire will stamp his good qualities on a large percentage of the calves. Therefore, the key to success with this method is in the selection of purebred bulls.

Fig. 3.2—Grading young bulls. University of California Extension Service suggests herd improvement by selecting cattle that have weight for age, grade high, and are regular producers. Here Reuben Albaugh, Livestock Specialist, scores a young bull on a ranch participating in a herd improvement program.

Crossbreeding—A crossbred is the offspring resulting from the mating of different breeds. The benefits of crossbreeding fall into two general channels: first, increasing general thrift, vigor, and rate of gain through hybrid vigor; second, securing better adaption or conformation by combining the desirable characteristics of different breeds, for example, combining the polled condition of Aberdeen-Angus with the milking ability of Shorthorns. Recently a great deal has been done in combining the heat resistance of Brahmas with different breeds.

A further application of the crossbreeding principles is to utilize crossbred dams mated to a third breed.

Comparison of Growth Between Hereford and Hereford X Shorthorn Heifers

	Hereford pounds	Hereford X Shorthorn pounds	Difference pounds
Birth weight	74	77	3
Weaning weight	380	400	20
Weight at 18 months	725	776	51
Weight at 30 months	1,050	1,138	88
Gain to 30 months	----	----	81

Source of data: Baker, A. L. and Quesenberry, J. R., 1944. Comparison of growth of Hereford and crossbreds. Journ. Anim. Sci., 3:322-25.

While excellent beef animals may be produced from this system, the herd segregation problem for most beef setups is too complicated for the results obtained. Indiscriminate crossbreeding can be dangerous, as the success of the program depends upon good foundation herds and consistent use of superior sires.

Therefore it becomes apparent there are both advantages and disadvantages to crossbreeding. Circular 543 of the California Agricultural Experiment Station enumerates these differences:

Advantages of Crossbreeding

1. Higher conception rate
2. Higher percentage of calf crop
3. Heavier calves at weaning
4. Earlier sexual maturity
5. Fewer calving problems
6. Higher milk production
7. Greater longevity
8. Greater insect and/or climatic adaptability
9. Fewer services per conception
10. Earlier conception in the breeding season

Disadvantages of crossbreeding

1. Loss of uniformity in color—esthetic value (in some crosses this can be avoided—Red Angus, Red Shorthorn, and Hereford)
2. Expense of maintaining and managing at least two herds
3. Usually not practical under unfenced range conditions

Crisscrossing and Rotational Crossing

These plans should produce the best results. The plans are: (1) two breeds—crisscrossing; (2) three breeds—rotational crossing.

Two-Breed Cross (crisscrossing)

Mate all Hereford cows to Shorthorn bulls for three calf crops and replace one-third of the old cows with crossbred yearling heifers each year. Identify each cow and divide the cows into two herds at breeding time.

Herd I	*Herd II*
Cows: Shorthorn-Hereford	Cows: Shorthorn-Hereford
breed only to Shorthorn bulls	breed only to Hereford bulls

← Heifer replacements ←

→ Heifer replacements →

Heifers sired by Shorthorn bulls are used for replacements in Herd II. Heifers sired by Hereford bulls are used for replacements for Herd I.

Herd I cows after several generations will standardize at $\frac{2}{3}$ Hereford and $\frac{1}{3}$ Shorthorn.

Herd II cows after several generations will standardize at $\frac{2}{3}$ Shorthorn and $\frac{1}{3}$ Hereford.

Three-Breed Cross (rotational crossing)

Mate all Hereford cows to Angus bulls for three calf crops and replace one-third of the old cows with crossbred yearling heifers each year.

Herd I	*Herd II*	*Herd III*
Cows: Angus-Hereford breed only to Angus bulls	Cows: Angus-Hereford breed only to Hereford bulls.	Cows: Angus-Hereford breed only to Shorthorn bulls

Heifer replacements (Herd I → Herd II)   Heifer replacements (Herd II → Herd III)

Heifer replacements (Herd III → Herd I)

Heifers sired by Angus bulls go into Herd II. Heifers sired by Hereford bulls go into Herd III and heifers sired by Shorthorn bulls are used for replacements in Herd I.

Cow herds will standardize in breeding as follows:

Herd I Cows	*Herd II Cows*	*Herd III Cows*
$\frac{4}{7}$ Shorthorn	$\frac{4}{7}$ Angus	$\frac{4}{7}$ Hereford
$\frac{2}{7}$ Hereford	$\frac{2}{7}$ Shorthorn	$\frac{2}{7}$ Angus
$\frac{1}{7}$ Angus	$\frac{1}{7}$ Hereford	$\frac{1}{7}$ Shorthorn

Need for Purebred Bulls

Superior bulls systematically used will improve the herd regardless of the quality of the cows. Therefore the purebred breeder will always be an important factor in meeting the need for high performance sires.

Purebred breeders of registered beef cattle always

Courtesy, American Hereford Association

Fig. 3.3—A purebred cow and her calf. Note how all characteristics have bred true and the calf is a replica of her dam in color markings, conformation, and so on. However, if the cow is genetically pure other important characteristics such as rate of gain, feed conversion, disposition, and other important profit factors will also be passed on from dam to offspring. Using the breeding methods outlined in this chapter, pure-bred breeders hope to achieve these characteristics in their animals.

mate animals of the same breed and operate on the principle of "like begets like."

Purebred breeders use the following methods:

Outcrossing—Outcrossing is the mating of relatively unrelated animals within the same breed; those having

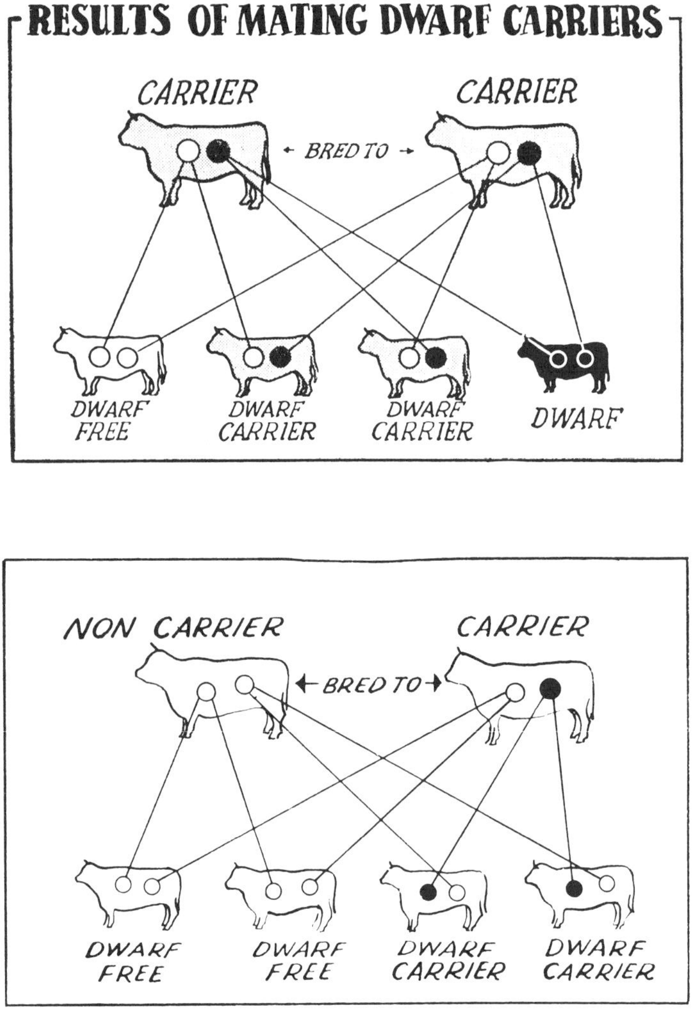

From Oregon Extension Bul. 743

Fig. 3.4—Compare the lower illustration with the one above. The important fact to observe is that for a dwarf to appear, both parents must carry the factor of dwarfism. Therefore, any culling program must consider this, especially in selecting replacement animals.

no common ancestors within the first four to six generations.

Inbreeding—Inbreeding is the mating of related animals; those having one or more common ancestors in the first few generations. Inbreeding is divided into two parts—closebreeding and linebreeding, as shown below. Linebreeding is the most often recommended.

Closebreeding—Closebreeding is the mating of sire to daughter, son to dam, full brother to sister.

Linebreeding—Linebreeding is the mating of half brother and sister or of those more distantly related, such as cousins. It was by such methods as these that the well-known beef breeds of superior type of today have evolved.

Regardless of the system of breeding used, results

Fig. 3.5—A dwarf calf. Characteristics include upturned muzzle, undershot jaw, potbellied barrel in addition to stunted size. Note appearance of dwarfs in several breeds as dwarfs appear in all common breeds of cattle.

should always be carefully analyzed in the offspring. The goal of all producers should be to breed out undesirable characteristics and at the same time retain the desirable qualities.

Cull Rigidly

In any system, the key to success is rigid culling. Cows that fail to breed or that raise poor calves of undesirable type should be culled as well as the bull siring heifers that are not an improvement over their dams.

Sell Any Cow Not Pregnant—This means, of course, sell any cow not pregnant when she should be. A desirable practice is to pregnancy test at weaning time, for by this period the cow should be bred back and be with calf for the next calving period. In a registered herd where the cow is very valuable and she is average or above average for the herd, she might be kept for one more year.

Cull Problem Heifers—Females that are shy breeders, difficult to get with calf, have irregular births such as hiplock, or have serious management problems, should be culled.

Eliminate Dwarfs

Breeders that are getting dwarfs will want to take steps to eliminate dwarfism from their herds. This calls for the eventual elimination of dwarf-producing bulls, and the use of dwarf-free bulls.

2. Keep Breeding Herd the Proper Size

Determine Number Head by Amount of Feed—Maximum profits are possible only when grass and feed are abundant. The number of cattle on any ranch should be determined by the amount of available feed. Percentage of calf crop is to a large extent dependent on maintaining adequate nutrition for the breeding herd. Cows kept in thrifty condition through adequate feeding will breed readily and produce a large percentage of calves uniform in age. In addition, the breeding herd should be continually culled to keep its numbers in line with the amount of feed the ranch is capable of producing. Non-breeders and otherwise unsatisfactory breeds, and cows that have outlived their usefulness as well as heifers that do not conform to the type level of the herd, should be sold. Culling may be done in the fall to obviate the additional expense of maintaining undesirable cattle throughout the winter.

Select Desirable Replacements—In selecting replacement heifers, consideration should be given to their capacities to consume feed and their respective weights at one year of age.

3. Select Replacements on a Performance Record

There was a time when the eye or visual inspection alone was sufficient to select replacement animals. However, most livestock has been improved enough that appearance by itself is not adequate to select replacements. For example, it would be difficult to select calves on their rate of gain simply by looking at them. Therefore, use of performance records is an approved practice in selecting replacements. The Pennsylvania State University (Progress Report 277) reports high estimates of heritability and genetic correlations found for daily gain and grade codes indicate selection can be effective.

Performance records from 2,019 Angus and Hereford calves located on 37 farms enrolled in the Pennsylvania Beef Cattle Improvement Program were used in this study, to (1) determine the average daily gain and grade by sex, within breed; (2) estimate the degree of relationship between these traits; (3) establish estimates of heritability and genetic, environmental, and phenotypic correlations; and (4) develop adjustment factors for average daily gain which take into consideration sex of calf and age of dam. From this work, new correction factors for sex and for age of dam were determined, and a new Weaning Index was developed for Pennsylvania.

Significant correlation coefficients were found between all the weaning traits—average daily gain, grade code, adjusted daily gain, and weaning index. These correlations were calculated within breed; however, corresponding coefficients for both breeds were found to be similar in all instances.

4. Breed Heifers at the Right Age

Breed According to Size—Growth and development rather than age are the best criteria for proper time of breeding heifers. Heifers that are bred too small

Fig. 3.6—Good quality replacement heifers. Consideration should be given to their weights at one year of age.

are usually stunted, never attain their full size and capacity, and produce poorly nourished, under-sized calves. There may be a heavy death loss of cows and calves at calving.

For well-grown heifers in good condition, breeding at 18 months and calving at 27 months would be ideal. This practice is common in many purebred herds although breeding at two years of age might be a safer practice. For commercial cattle, breeding at two years of age and calving first at three years is a standard practice.

Heifer calves should weigh 450 pounds or more when weaned, should be fed for normal growth (1 pound or more daily) from then to the next grass season, and should weigh 650 to 700 pounds when bred.

Breed Yearling Heifers

On many ranches greater economic gains are possible if heifers are bred as yearlings providing certain precautions are taken. University of California Circular 433 outlines the following program for best results:

Breed only thrifty high-grading heifers that weigh 600 pounds or more at breeding time.
Feed for continuous growth.
Give special care at calving time.
Veal their calves at 3½ months of age.
Use young, small-bodied, small-boned bulls.

Breed only a few heifers the first year of the program—allow the operator to gain experience in managing the project. The same is true of any beef-cattle-improvement program where the human element is closely involved.

Breed for Small Calves—Successful calving of heifers bred as yearlings can best be accomplished by using

Fig. 3.7—Heifers should be well grown out, like these Shorthorns, before breeding.

bulls which tend to produce small calves. Many of the beef breeds, especially Hereford and Angus, produce suitable calves. Individual differences are also important as some bulls may be small boned or small themselves and have a history of producing small calves. These kinds of bulls should be used; under no conditions use a bull such as a Holstein or those known to produce large calves simply because they are on hand. Angus bulls are often used because the calves are small and polled, and because young bulls are easily available.

5. Managing for Artificial Insemination

In recent years artificial insemination (A.I.) has become a tool used frequently by cattlemen even though it means changing their methods of operation. Breed associations have lowered restrictions against it,

Fig. 3.8—The first-calf heifers and their healthy crossbred calves. These heifers, mated as yearlings to an Angus bull, calved unaided and with little difficulty, as the calves were small. The calves will be sold at 3½ to 4 months of age in order that their dams will not be retarded in their growth.

crossbreeding is made more practical, and ranchers who are performance minded can obtain the best germ plasm economically. This method may not be the answer for everyone so each rancher should carefully evaluate his own situation. University of California Leaflet 2317 discusses this practice in detail. According to this leaflet, the producer must have the proper attitude and a determination to make A.I. work. He should become knowledgeable about the techniques and routines involved, consult authorities, and read. A.I. schools are available. Increased labor is inevitable. Improved facilities are essential. A short breeding season is vital from 25 to 30 days. Proper nutrition is essential. Heat detection is critical.

6. Maintain a High Percentage Calf Crop

This means the number of calves weaned out of

every 100 cows bred. The greater this percentage, the lower will be the cost of production. Good managers will have as high as 90 to 95 per cent calf crop.

An average of seven years' data, on the San Joaquin Experimental Range, University of California, comparing supplemental feeding with no supplemental feeding is as follows:

	Group A Fed Supplements	Group B Not Fed Supplements
Percentage of pregnancies	89.6	77.2
Percentage calf crop weaned	83.0	67.6
Average weaning weight, pounds	470.0	417.0
Average calf production per breeding cow, pounds	390.0	282.0
Average yearly supplemental feed, pounds	365.0	18.0

Cull Dry First-Calf Heifers—One of the best ways to insure a high percentage calf crop is to eliminate "free loaders" or cows who have an irregular calving record. Much of this can be accomplished by culling at first-calf heifer time. According to a Montana study, heifers that were dry when they were supposed to have their first calf had a lifetime record of only 55 per cent calf production. Heifers that dropped a calf the first year averaged 87 per cent lifetime calf production.

7. Select the Proper Breeding Season

Breed at Proper Time—One of the most important items to consider is the time of breeding and consequent calving season. From a market standpoint, it is desirable to have all cattle as near uniform size and condition as possible. In purebred herds where a show herd is developed, it is desirable to have the calves well distributed throughout the year. However, if such a herd is specializing in bulls for the range trade, cows should be bred to drop fall or early winter calves. In

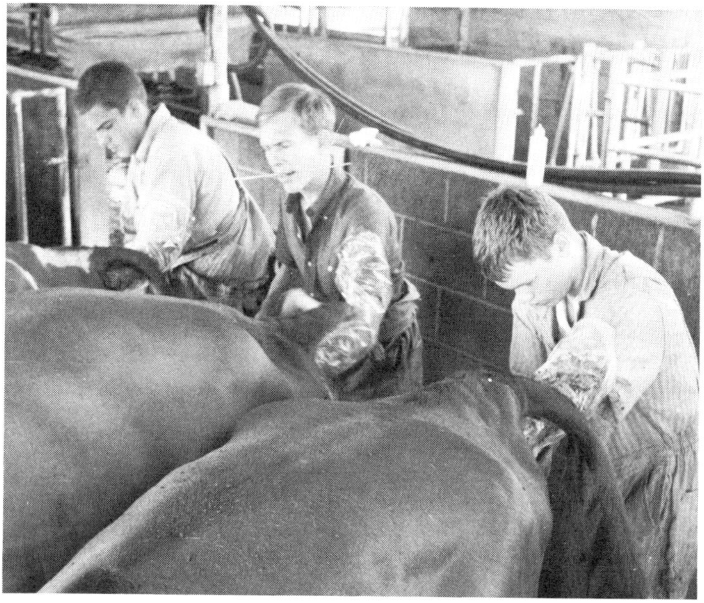

Fig. 3.9—Students learning artificial insemination at a community college. Many cattlemen are learning to perform such practices as diagnosing pregnancy and administering artificial insemination. Both of these approved practices can do much to increase the percentage of calf crop. They may also assist in bunching calves so they are dropped close together and at a desirable period.

mild climates, it is desirable to have calves dropped in November, December, or sooner.

Maintain a Short Breeding Season—A short breeding season of around three months is desirable so there will be no "off-season" calves and the calf crop will be uniform in age and size. The calf crop should be timed to come as early as possible and yet escape the danger of late storms. In mountain areas, bulls should be placed with the herd in June or July, but in valleys or warmer areas, the breeding season can start several months sooner.

Heat periods will normally recur about six weeks after calving. Then the heat periods generally will occur every 19 to 21 days until the cow is settled.

8. Use the Right Number of Cows per Bull

The most successful cattlemen permit the bulls to run with the cows only during the breeding season. Bull calves should not be put into service until they are at least fourteen months of age, and then only on a limited number of young cows.

Get Bull Acclimated—Bulls should be acclimated and in good condition but not fat when put into service. In purebred herds or in a small field of good pasture, a

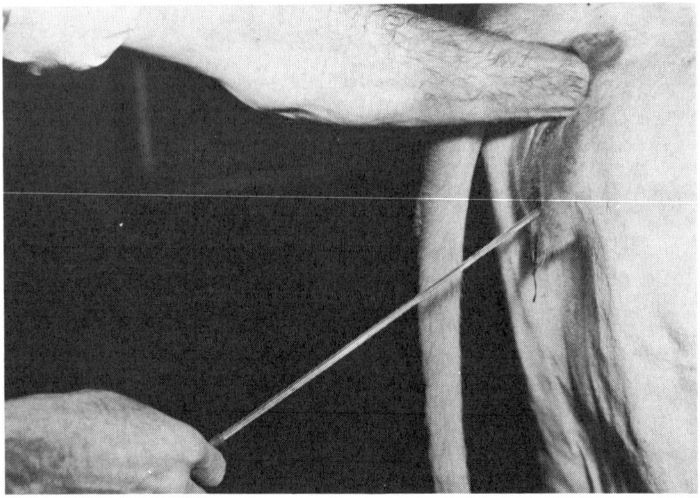

Fig. 3.10—Properly inserting the pipette is one of the first steps in artifical insemination (A.I.). This method of breeding can be used to advantage in growing beef cattle in confinement, such as in part-time farming or drylot cow-calf operations. Some range operators consider it a desirable practice, although there are obvious hurdles under range conditions.

Courtesy, American Hereford Assoc.

Fig. 3.11—With desirable bulls like this one best results are obtained by rotating bulls every two weeks in the breeding herd.

mature bull will serve as many as fifty or seventy-five cows. On range conditions or in very large pastures, *three or four bulls* are recommended *for every 100* cows. Bulls two years of age should not be put on the range.

Rotate Bulls—On fenced ranges, "rotation of bulls" is a desirable practice. Under this system, half the bulls are turned in with the cows for ten to fourteen days; then these are taken out for rest and extra feeding and fresh bulls turned in with the herd. After two such shifts, all the bulls can be left with the herd for the remainder of the season.

In a purebred herd which has an extremely valuable sire, artificial insemination would be one way of increasing the number of calves sired by one desirable bull.

Fig. 3.12—Good fences will greatly facilitate separating animals into groups. Native materials as shown in the fence above make attractive and economical fencing.

9. Care for the Herd Bull Properly

A bull's worth depends on his performance. Proper care will tend to insure the right kind of performance. Healthy, vigorous bulls sire more calves than do weak, thin bulls.

Winter Bulls Separately—Most bulls lose weight during the summer season if allowed to run with the cows, and need to gain weight during the winter. The herd bull should be wintered separately from the cows in a pasture or in a lot large enough to allow for exercise.

Bulls kept on the range that are in poor condition in the spring will be unfit for service. For herd bulls, the following rations may be used, according to Florida Extension Bulletin 104:

Ration 1—Two to three pounds cottonseed cake on good pasture.

Ration 2—Two to three pounds cottonseed cake, fifteen to twenty pounds silage or ten to twelve pounds native hay.

Ration 3—Fifteen to twenty pounds legume hay.

Feed any one of the above rations depending on availability of feeds. On a good legume type irrigated pasture, little or no supplemental feeding is necessary.

Do Not Get Bulls Too Fat—A bull should not be fat or he will become a slow breeder. Exercise is of the utmost importance, especially in purebred herds where bulls are not on the range.

The feet should be trimmed as often as necessary to keep them level and the toes open. Poor feet make a poor breeding bull. Flies bother the animals greatly, especially the older bulls. Regular spraying with an insecticide is a desirable practice.

If you use approved practices, a good bull should give service for six or seven years. Be prepared to replace a valuable sire. According to American Hereford Association, replace a valuable bull by obtaining a young bull to use on the heifers. Save a son of the old bull and try him on the cows before selling the old bull.

10. Care for the Breeding Herd Properly All Year

Regardless of the size or class of the beef cattle operations, there must be a complete yearly cycle of feed and pasture. Plenty of water and salt should be available at all times. The proper mineral supplement is essential in areas that have serious soil deficiencies. Ample exercise, while no problem on the range, is important for the health of cattle when confined in small enclosures.

Divide Animals into Age Groups—It is desirable to divide the females into age groups—the cows in one group and the heifers in another.

Cows Must Gain Weight—While cows do not have to

be fat in order to insure maximum usefulness, they should be maintained in good, strong shape. It is highly desirable that a cow gain in weight between the time of weaning and calving. The gain in weight should be at least equivalent to the amount lost in producing her calf.

Managing the Herd in Summer—Summer management of the cows is most important. Calves should make rapid and economical gains on the increased milk flow of cows on good pastures. Plenty of water and shade pay big dividends. Make certain water tanks are not too tall for calves to drink from. If the pasture is poor, calves should be creep fed in order to insure proper growth and maintain health of the dam. Regular inspection of the herd is a must, especially if screwworms are a problem.

Managing the Herd in Winter—Animals should be grouped by ages for winter feeding and care. All females not nursing calves should be kept out of doors providing they have access to a well-constructed open shed facing in the opposite direction of prevailing storms. Cattle can stand cold weather if they are away from draughts. In mild climates, rolling hills constitute a desirable place to winter cattle as they can get on the lee side of hills for shelter.

According to Farmer's Bulletin 1916, it is desirable to feed fodder in pastures that need fertilizer so that both cattle and pasture will benefit.

Suggested rations for wintering dry pregnant cows are:

		Pounds	
Ration 1—	Corn or sorgo silage	35	U.S.D.A. 1916, p. 15
	Legume hay	6	
Ration 2—	Alfalfa hay	20	Calif. 131, p. 98
Ration 3—	Protein supplement	1½ to 4	
	Dry grass	At will	

Some forage of bright green color is recommended to insure against Vitamin A deficiency.

A good general rule is to feed two to three pounds of hay per hundred weight per day or substitute three pounds silage for each pound of hay up to half the daily feed of hay.

Feed Cows Early in Fall—It is desirable to begin to feed cows early enough in the fall to keep them from losing too much weight.

11. Keep a Record of Performance

Adequate records are essential in order to measure improvements in breeding beef cattle. The first step in applying a breeding program is to determine the qualities of the ideal type. The second step is to evaluate the individual animals and the third is to select and mate superior stock.

Identify Each Animal—The first step in any improvement program using records of performance is that each animal must be identified by number. Rate of gain or other evaluative criteria can be used only when each individual animal can be positively identified. It is essential that this system be permanent and if possible easily and quickly read. Many systems are used such as neck chains, horn brands, tattoo or hair clipping. However, in most cases, the practical way is branding.

For range cattle neck chains are dangerous as the chain may get caught in a crack in the rocks and cattle strangle. Horn brands are satisfactory assuming the cattle in question have horns. If hide brands are used an identification number can be branded just below the brand and thus be easily seen. This system is used quite often. Tattooing is satisfactory except that each animal must be caught and examined which takes considerable time and effort. However if properly equipped with chutes, etc., the task can be made easier.

Clip Before Branding—If the hair is short as when animals have just shed out, clipping may be unnecessary. However, long shaggy hair tends to prevent a uniform brand and makes it difficult to determine just how long to apply the hot iron.

12. Determine the Stage of Production to Enter

According to Shultis and Strong, California Extension Service, consideration should be given to the three stages of beef production.

Three breeding methods may be used:

1. Mass selection, which consists of choosing the best animals by appearance without knowledge of ancestry. Used extensively in commercial herds.

Fig. 3.13—Ideal type Shorthorn female. Only by keeping records of performance is it possible to upgrade herds toward the type exhibited by this animal.

2. Pedigree breeding, commonly used in purebred herds, consists of a knowledge of ancestry plus individual appearance.

3. Determine transmitting ability of animals and mate accordingly. To determine the transmitting ability of an animal, progeny performance records are kept as follows:

a. Weight for age, to measure quantity, e.g., W.D.A.

b. Over-all evaluation by grading indicates degree of excellence. Although this method has limitations, it serves as a relatively precise index.

Use Records to Make Decisions

The following indicates how a commercial herd may be improved by selective breeding based on a 100-cow herd with a 20 per cent replacement per year:

1. Pick out the top females of breeding age for the select herd.
2. Weigh and grade cows in selected herd.
3. Breed this herd in a separate field.
4. Use the best bulls you have available with the select herd.
5. Where possible select bulls for rate and efficiency of gain, as well as type, conformation, and pedigree.
6. At weaning, cull from the select herd those cows that fail to produce good-type growthy calves.
7. Also at weaning, weigh and grade the calves.
8. Make first selection of replacement heifers when they are weaned.
9. Make final selection when heifers are ready to go into breeding herd.
10. Top heifers should go to the select herd; others in the general herd.

11. The select herd may be increased in size to produce all replacements for both herds.

Select Stage Suited to Your Farm

• In the first stage, the animal is a weaner calf in a breeding herd, usually on a range cattle ranch.

• In the second stage, the calf is raised to the yearling-feeder stage. This may be done on the same ranch or elsewhere, perhaps on irrigated pasture.

• In the third stage, the feeder is finished to a grade of *good* or *choice* for slaughter.

Compare Alternatives

Compare alternative plans under expected price and cost conditions to determine quickly and surely the most profitable way to use feed in beef production. Trial and error is slow and hazardous. Even if you develop a profitable production program, you must anticipate the need to change with changes in price differentials or feed supply. A yearly trial budget improves your chances for a good profit.

13. Bulls vs. Steers (for beef)

For many years foreign countries have been marketing beef without discriminating between bulls and steers. In this country according to USDA grading data, only a tiny percentage of commercial beef is of bullock grade. According to Leaflet 2310, University of California, Davis, several studies emphasize the need to revise our thinking about producing and marketing young bull beef. Summarized, these studies show:

• Bulls gain faster and more efficiently under full feeding than do steers of the same age and quality.

- Under similar pasture conditions, steers may show greater rates of live gain than do bulls.
- In most trials, late castrates (7 to 8 months old), made faster preweaning but not postweaning gains than did early castrates (approximately 6 weeks old).
- No management problems are reported when finishing young bulls of the beef breeds when they are not over 15 months old.
- Bulls produce more red meat in the carcass on both trimmed and untrimmed basis than do steers.
- At the same age and under the same environment as steers, bulls' carcasses tend to have a somewhat lower USDA quality grade and, therefore, a lower final grade, mainly due to less marbling.
- Implanting bulls with diethylstilbestrol* at a young age improved the rate of gain in some trials but not in others. Carcass quality was improved by implantation. Other implants are available.
- Some taste panels found that young steer meat was somewhat superior to bull meat in flavor and tenderness (steers and bulls both were about 15 months old). Other panels found no difference. In all cases, both products were acceptable.
- Since bull beef is acceptable to taste panels and consumers and the lean yield from these carcasses is high, serious consideration should be given to the establishment of USDA carcass grading standards for young bulls that are interchangeable with those for steers and heifers. This would greatly aid in the marketing of this desirable product.
- Beef from short-scrotum bulls was found to be as acceptable to the consumer in taste tests as steer beef.
- In a U.C. study, short-scrotum bulls outperformed intact bulls and steers in weaning and post-weaning

*Currently banned.

performance. They produced the heaviest and leanest carcasses and were the most efficient in feedlot gains.

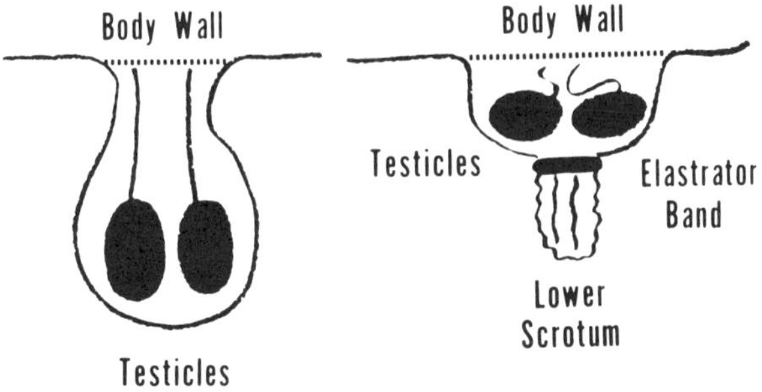

Fig. 3.14—Short-scrotum method. In this operation the testicles are pushed up close to the abdomen and a rubber band is placed around the scrotum beneath the testicles. This renders animals sterile but allows the production of the hormone which affects growth.

Chapter IV

PRODUCTION OF CALVES AND FEEDER CATTLE

The percentage of calf crop, number of calves raised, and number of death losses have a more direct bearing on production costs than almost any other combination of production factors. The number of calves produced and proper development of the calves until they are weaned six or seven months later is to a large extent dependent upon the care the cow receives. Therefore, it is highly important that approved practices be used in handling the cow as well as in the development of calves.

Activities Which Involve Approved Practices

1. Caring for the cow during pregnancy.
2. Caring for the cow at calving time.
3. Caring for the newborn calf.
4. Caring for the cow after calving.
5. Keeping proper identification.
6. Reading brands.
7. Creep feeding calves.
8. Providing proper minerals.
9. Properly handling the calves.
10. Dehorning and castrating calves.

Figure 4.0—Babysitting. Here a mother cow is babysitting her own and half a dozen other calves while their dams graze some distance away. Such a phenomenon is common and generally indicates "all is well" in the herd.

11. Properly vaccinating calves.
12. Providing satisfactory pastures.
13. Properly weaning calves.
14. Producing feeder cattle.

1. Caring for the Cow During Pregnancy

Have Cows in Good Flesh Before Calving—The drain of gestation is relatively small and pregnancy may even stimulate growth. However, in order that the cow may properly nurse her calf in later months, it is necessary to get her in good flesh before calving. The cow should be fed a balanced ration which includes legume hay or pasture for at least three or four months before calving.

Supplement Range Cows—In a 12-year study of effects of supplementing the diets of range cows Bulletin

GESTATION TABLE FOR COWS (283 DAYS)

Day of month bred	Jan.	Feb.	Mar.	Apr.	May	June	July	Aug.	Sept.	Oct.	Nov.	Dec.
	Explanation: Find date cow was bred in first column and month bred in top line. The date in column below opposite date bred will be the time at which the cow is due to calve											
	Oct.	Nov.	Dec.	Jan.	Feb.	Mar.	Apr.	May	June	July	Aug.	Sept.
1	11	11	9	9	8	11	10	11	11	11	11	10
2	12	12	10	10	9	12	11	12	12	12	12	11
3	13	13	11	11	10	13	12	13	13	13	13	12
4	14	14	12	12	11	14	13	14	14	14	14	13
5	15	15	13	13	12	15	14	15	15	15	15	14
6	16	16	14	14	13	16	15	16	16	16	16	15
7	17	17	15	15	14	17	16	17	17	17	17	16
8	18	18	16	16	15	18	17	18	18	18	18	17
9	19	19	17	17	16	19	18	19	19	19	19	18
10	20	20	18	18	17	20	19	20	20	20	20	19
11	21	21	19	19	18	21	20	21	21	21	21	20
12	22	22	20	20	19	22	21	22	22	22	22	21
13	23	23	21	21	20	23	22	23	23	23	23	22
14	24	24	22	22	21	24	23	24	24	24	24	23
15	25	25	23	23	22	25	24	25	25	25	25	24
16	26	26	24	24	23	26	25	26	26	26	26	25
17	27	27	25	25	24	27	26	27	27	27	27	26
18	28	28	26	26	25	28	27	28	28	28	28	27
19	29	29	27	27	26	29	28	29	29	29	29	28
20	30	30	28	28	27	30	29	30	30	30	30	29
21	31	Dec. 1	29	29	28	31	30	31	July 1	31	31	30
22	Nov. 1	2	30	30	Mar. 1	Apr. 1	May 1	June 1	2	Aug. 1	Sept. 1	Oct. 1
23	2	3	31	31	2	2	2	2	3	2	2	2
24	3	4	Jan. 1	Feb. 1	3	3	3	3	4	3	3	3
25	4	5	2	2	4	4	4	4	5	4	4	4
26	5	6	3	3	5	5	5	5	6	5	5	5
27	6	7	4	4	6	6	6	6	7	6	6	6
28	7	8	5	5	7	7	7	7	8	7	7	7
29	8	......	6	6	8	8	8	8	9	8	8	8
30	9	......	7	7	9	9	9	9	10	9	9	9
31	10	......	8	......	10	......	10	10	......	10	......	10

From U.S.D.A. Farmer's Bulletin 1916.

822 of the California Agricultural Experiment Station reports that supplemented herds averaged 10.5 per cent dry cows, 6.0 per cent loss from abortions and still-births, and 3.2 per cent loss from dystocia, death of dam, and other causes; unsupplemented cows averaged 25.0, 5.0, and 7.0 per cent, respectively. The excessive numbers of dry cows in each group may largely be due to the effects of close grazing on one-third of the cows in each group, plus the lack of needed supplements in the unsupplemented group. A 10-year investigative period which followed the first study of the effects of climatic conditions on weight gains and losses indicated that the rates of protein and energy supplementation used in the program were adequate for good conception. Pregnant beef cows can utilize rice straw, barley

Courtesy, Dow Chemical Company

Fig. 4.1—Range cattle fed a supplement have fewer dry cows each year than those not fed a supplement. These cows are being supplemented with Kedlor, a non-protein nitrogenous feed with a complete mineral mix. This supplement permits cows to utilize high roughage cornstalks efficiently and to maintain weight and produce healthier calves.

straw, and other low-cost feeds if adequate protein, vitamin, and mineral requirements are met.

Separate the Pregnant Cows from Other Cattle— As the cow approaches parturition her temperament changes and she becomes nervous and likes to be alone. Therefore, it is best if the cows about to calve are separated from the other cattle so they can remain quiet and will not be jarred or jostled by other animals milling around.

Provide a Suitable Place in Which to Calve—If the weather is suitable, a well-drained pasture is an ideal place in which to calve. Pastures on rolling hills or natural windbreaks will be a better place. In extreme weather, shelter should be provided such as a well-bedded box stall about 10 x 12 feet. If the cow is to calve in a stall, she should be put in several weeks beforehand to get accustomed to these surroundings.

Utilize Natural Features of Pasture—Pastures with good drainage, wind breaks, and warm exposures are good places to calve. Year-around grazing in one pasture is not necessarily the cause of slower rebreeding, lower conception, higher reproduction losses, and lighter calfweaning weights. Increased production may be obtained by making better use of favorable natural features, such as those which exist in pastures having a steeper slope and southern exposure, resulting in a warmer area and earlier forage.

Provide Ample Fresh Clean Water at All Times—This will assist in maintaining proper physiological processes and aid in general health. In winter it should be unchilled.

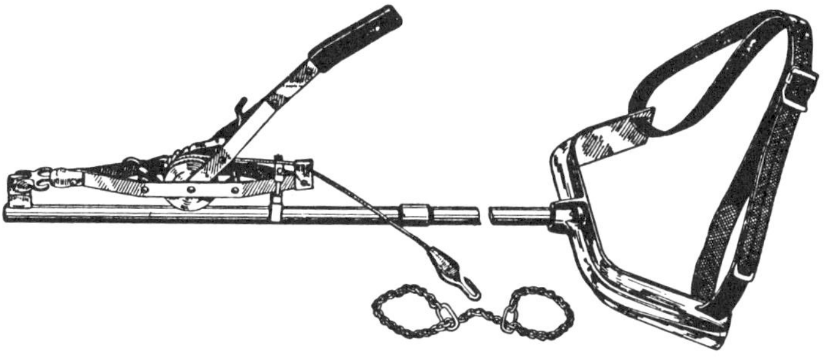

Courtesy, Franklin Serum Co., Denver, Colo.

Fig. 4.2—A mechanical calf puller. This item is an important piece of equipment on a cattle ranch. However it should be used only when necessary and precisely as manufacturer's directions specify. Most cows calve unaided.

2. Caring for the Cow at Calving Time

Let the Cow Become Accustomed to Her Surroundings—In order that the cow be settled and quiet, she should feel at home in her surroundings. If the herd must be moved to a new pasture or the pregnant cows separated, it is well to do so several weeks or more before calving in order that they can explore their new home and satisfy their curiosity. This is especially important if the animals must be kept in a barn because of foul weather or if they are to be in individual stalls as practiced in many purebred herds.

Reduce Feed—If the cow is being fed grain, reduce the amount being fed or omit altogether as parturition

Fig. 4.3—Mother cow licking her newborn calf. Large majority of cows calve successfully unaided. They should be watched but not bothered.

approaches. During this period a good feed to include in the ration is bran.

Be on Hand to Render Assistance—Cows calving in spring and summer on good pasture seldom need assistance during parturition. However, heifers or cows whose gestation period has been unusually long should be closely watched. If it is obvious the cow is in trouble, do not wait until she has worn herself out before rendering aid. Oftentimes by placing a small cord around the front feet of the calf or by grasping the feet with the hands and pulling the calf toward the cow's hocks, an attendant can assist the cow in delivery of the calf. It is important to pull only when the cow is straining and pull only toward the hocks of the cow; not straight away from the cow.

Mechanical calf pullers are available commercially. If properly used they are important items on a ranch.

Caution must be taken to see that the calf is in correct position for delivery.

Do Not Give Too Much Attention—Never assist the cow as soon as labor movements begin. Experienced stockmen realize the importance of this warning, but the natural thing for an inexperienced person to do is to try and help. A good attendant is one who keeps his "eye" on developments, but does not make himself obvious to the cow. Too much attention worries the cow. She becomes nervous and moves around. Remember, the vast majority of range cows calve successfully unaided. Even in purebred herds visitors should be kept away from the cow when she is about to calve.

Call Veterinarian if Necessary—This should be done if birth is delayed more than two hours or if the calf is in an abnormal position. Normally the calf's front legs appear first with the head resting on those legs. It is also normal if both hind legs come first. If the tail appears first, the calf is coming backwards and sitting down. Call the veterinarian immediately. In commercial herds oftentimes all assistance is from the owner or attendant.

3. Caring for the Newborn Calf

Make Certain the Calf Is Breathing—If the calf seems lifeless or very weak at birth, examine the nostrils and mouth and remove any mucus that may be obstructing normal breathing. Sometimes blowing air into the mouth and nostrils after removing the mucus

Fig. 4.4—Disinfecting the navel of a newborn calf.

may aid in reviving the calf. Should it appear to have liquid in its throat or lungs, the calf may be held upside down by its hind legs for a few moments.

Be Sure the Calf Gets the "First Milk"—The "first milk" or colostrum milk of the cow is a thick yellowish milk. It is important that the calf receives some of this milk in order to insure proper health of the animal. Small and weak calves should be assisted in nursing through the first day after birth.

Disinfect the Navel—The newborn calf should have the navel cord painted with tincture of iodine in order to prevent any possible navel infection. This operation is not often done with cattle on the range, but is especially important if calves are dropped around the barnyard or in permanent lots.

Prevent Fly Strike—If the weather is warm and there is danger of screwworm infestation, the area around the navel or near open wounds should be smeared with a fly repellent. Pine tar oil or any of the modern strike repellent smears are generally effective.

Keep the Calf Warm—If calves are dropped in spring or summer, there is no problem. However, "off-season" calves or those in cold climates should be well-bedded and sheltered during bad weather until they are husky enough to withstand the elements. Protection against cold winds is most important. Under these conditions, purebred breeders sometimes vaccinate against pneumonia.

Save the Orphan Calf—Occasionally a cow will die during parturition. Fast action can save the calf. If possible, squirt some of the first milk from the mother's udder into the mouth of the calf. Get another cow who is a good milker to adopt the youngster for her own.

Many cows will fight strange calves away, but most herds have a few cows who will let calves other than their own nurse. A good suggestion is to have a dairy-type nurse cow on hand for just such emergencies.

4. Caring for the Cow After Calving

Remove the Afterbirth—The placenta, or afterbirth, should be expelled in, at most, 12 hours after calving. If it is retained longer than 24 hours, a veterinarian should be called to remove it. Under no circumstances should it be forcibly pulled away. If calving occurs near the farmstead, the afterbirth should be burned or buried as a good sanitary measure. Range producers, however, seldom, if ever, bother with disposal.

Feed Cows Lightly at First—The cows should be fed lightly the first two weeks after the calf is born because too much rich feed will cause an increased milk flow and the calf may scour or have indigestion.

Provide Adequate Pasture and Minerals—Good green pasture not only supplies a well-balanced diet for cow and calf, but is an excellent place to maintain health and reduce disease. Pastures should not be grazed below six to eight inches, as worm eggs and other parasites are more numerous close to the ground. Young animals are especially susceptible to parasitic infestation. Water and salt should be available at all times in any pasture.

Inspect Udders Regularly—Beef cattle seldom have numerous udder troubles as do dairy cattle. However, regular inspection of the udder immediately after calving is a good practice so as to spot and treat any difficulty.

Veal Calf from Heifers Bred Too Young—While

pregnancy is no strain on heifers, lactation will stunt their growth. Calves should be disposed of for veal at eight to ten weeks of age to stop the drain of lactation and permit normal development of the heifers.

5. Keeping Proper Identification

Keep a Record of Births—U.S.D.A. Farmer's Bulletin suggests the herdsman have two breeding books —one of pocket size for daily use, and a larger office book to record duplicate dates of all breeding and calving dates. Irregularities in breeding, calving, and gestation periods should be noted.

Brand Range Calves—Permanent identification by branding is a desirable practice for all commercial herds. To avoid unnecessary damage to hides and ani-

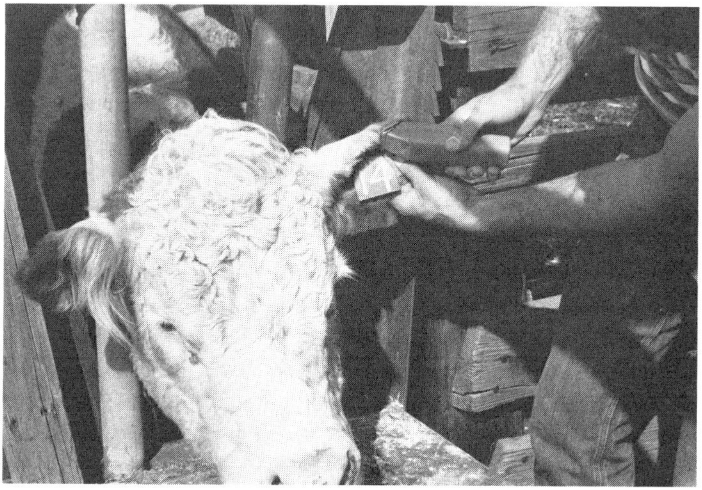

Fig. 4.5—Periodically trimming the long hairs on the ear may be necessary to facilitate fast, easy reading of the animal's I.D. number. Proper placement of the ear tag will assist in eliminating this chore on most animals.

Fig. 4.6—Branding calves with aid of tilting-type branding table. The hair of calves should be clipped prior to branding if the hair is too long or matted.

mals, brands should be as small as is consistent with ease of identification. Hot iron branding is most commonly used and in general is most satisfactory. Calves should be branded before they are four months old and while they are with their mothers. Branding by use of the "cold" method or caustic agents seldom gives a brand easy to read after the first year.

Horn branding is a desirable practice. As an easy way of quick identification, animals should not be branded until the animals are ready to be bred. Branding earlier will result in numbers chipping out because horns are still growing.

Heat Iron to Grey-hot Heat—The brand is applied

Fig. 4.7—Branding cattle with aid of squeeze chute. Iron and animal should be held firmly so the iron will not slide and cause the brand to become blotched. The brand should burn the skin to a light tan within a few seconds. Clipping the hair first may improve accuracy of brand location on the animal.

with a grey-hot iron, about the color of the branding fire ashes. A red-hot iron produces an over burned brand, often resulting in sores which may become infected. Cattle are never branded when hides are damp, as this causes a scalded or blotched mark. If the iron is not hot enough it will not produce a permanent mark and will make reading difficult.

Freeze Branding

In recent years an alternate method of branding

called freeze branding has been developed. This method uses a super-cooled iron rather than a hot one. However, it may not be legal in all states, and not all ranchers approve of this method—partly because it is quite involved and partly because they do not like the resulting brand.

The freezing destroys the melanocytes (pigment-producing cells). This causes the hair and the skin to turn white. Even white animals can be branded with this method as there is enough pigment in the skin of white animals to result in a white brand free of hair. Simply overbrand the white animal by leaving the iron on the skin longer than the required 30 seconds for other animals. Overbranding destroys the hair follicles.

The freeze-brand method has several advantages over the hot-iron method:

1. The brand is visible the year around without the hair having to be clipped. Freezing turns the hair white and removes the pigment from the skin.
2. The method is relatively painless to the animal. This is an important factor because of the public's concern for the health and safety of animals.
3. The method does not result in a thick scar which reduces the value of the hide as the hot-iron method does.
4. The brand does not blotch.

The procedure found to be most practical to date is as follows:

1. Clip the hair from the area to be branded. The closer it is clipped the better.
2. Dampen the area with alcohol to give better contact with the iron and to keep the iron from freezing to the hide.
3. Cool the iron by placing it in a solution of dry ice

(solid carbon dioxide) and alcohol (ethyl, methyl, or isopropyl). A styrofoam box (12 x 20 x 10 inches deep) is satisfactory as an insulated container. Do not use a plastic-lined box as thin plastic will crack. Break dry ice into egg-sized pieces to cover the bottom of the box. Add enough alcohol (95 per cent methyl) to cover the ice by 2 or 3 inches (any substance that will cool the iron may be used, such as liquid nitrogen; however, cost is an important factor to be considered). Leave the iron in the solution until heavy bubbling (or boiling) stops.

For quantity branding, it is desirable to have a large enough container to be able to place all of the irons being used in the solution at once.

4. Press the iron against the animal's skin for 30 seconds. A heavy copper or bronze branding iron (5/8 inch wide and at least 1 inch deep) has been proven to be best. If a smaller iron or an iron made of steel or iron (metals that will not replace the heat as quickly) is used, the branding will take longer. Also, a round-faced iron has been proven to be more efficient because it results in a more even contact with the skin.

Hair Brand for Short Periods

California Extension Service suggests hair brands to identify cattle for short periods. Designs are best made with an electric clipper and should be the width of the clipper blade and about 4 inches long.

Tattoo or Tag Calves in Purebred Herds—Most registry associations require tattoo identification on the inside of the ear, which is one of the most permanent means of identification. With breeds having pigmented skins, such as the Aberdeen-Angus, tattoo marks are

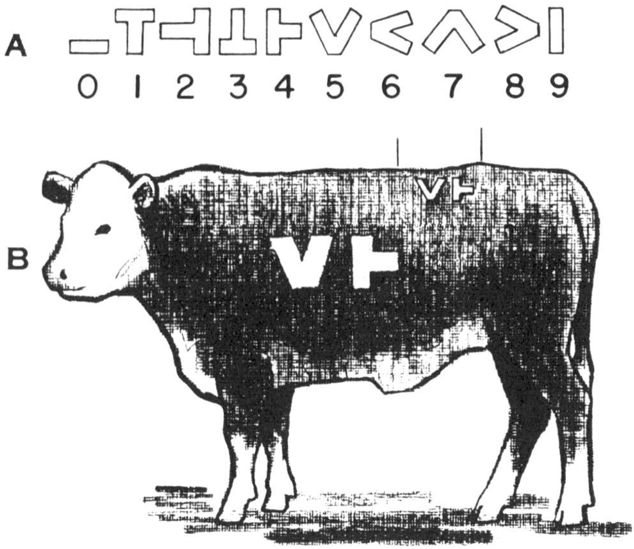

Fig. 4.8—A. Code for hair brands. B. Code 54. (Developed by Kenneth A. Wagnon, Specialist, Department of Animal Husbandry, Davis.)

difficult to read. Neck chains with numbers are used with Polled breeds. Ear tags are also used, but are sometimes torn out unless carefully inserted. In tattooing, the ear should be well-cleaned in the area where the mark is to be put and only permanent type tattoo ink used. The American Polled Hereford Association gives the following tips for good tattooing.

1. Wipe both ears clean.
2. Place tattoo between ribs of the ear.
3. After ear is pierced, rub tattoo ink into the punctures with thumb or finger, using plenty of ink.
4. Be sure the number in your herdbook corresponds with the numbers and letters in the ears.

In average herds, it is best to start with number 1 in each ear and proceed upward in regular order to 999.

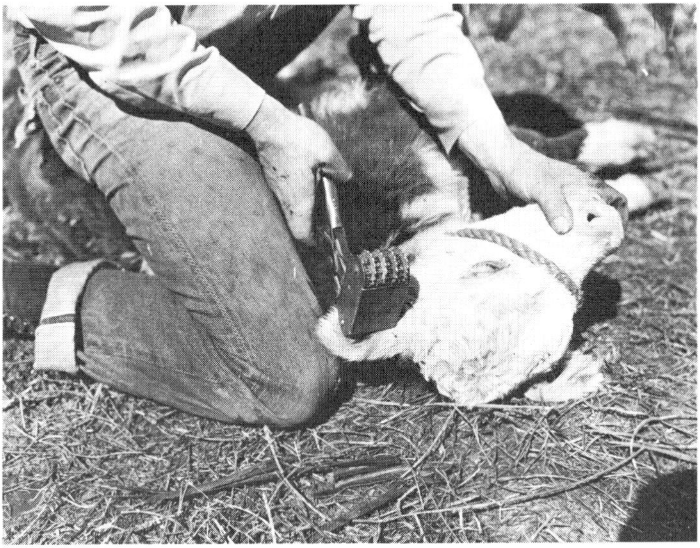

Fig. 4.9—Tattooing calves for permanent identification.

Properly Identify Calves and Dams—In working range cattle or large purebred herds, it is difficult to accurately determine which calf belongs to what dam. Separate cows and calves for two or three hours until the calves are hungry; then let one or two calves into the bunch of cows and they will quickly find their mothers. Try the following approved practice.

Glue on Numbered Tags to "Pair Up"

Even though calves and their dams may be ear tattooed, it is difficult to pair them up in order to record the identification in the record book or to apply for registration papers. An easy system is to duplicate numbered tags and glue them on the animals, one on the dam and one on the calf. Once they have been paired up and the data noted, the tattoo number can

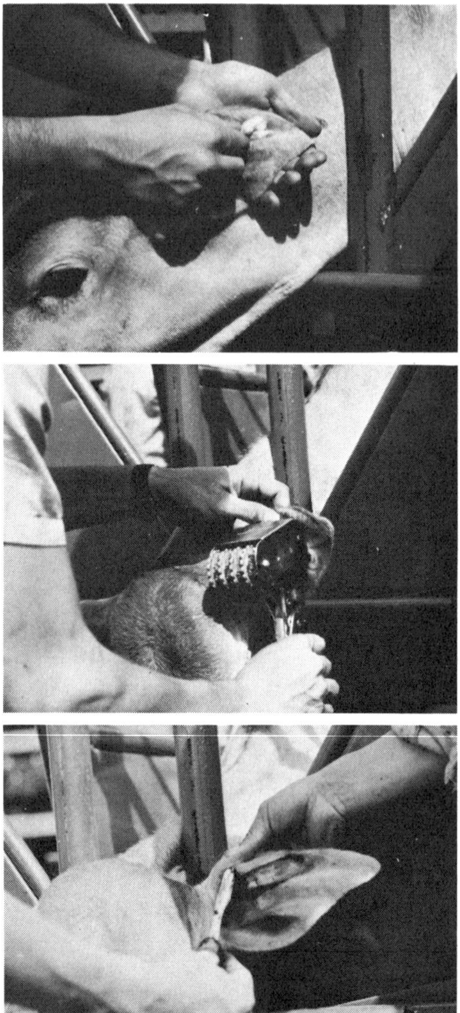

Fig. 4.8—Three steps in properly tattooing a calf: (Top) clean ear with cloth moistened with alcohol; (center) squeeze tattoo instrument firmly so skin is punctured; and (bottom) brush in tattoo ink briskly with old toothbrush.

be found and proper I.D. made. If the dam is already legibly tagged or identified, only the calf need have a number glued on. Excellent glue for this purpose is available at any livestock supply house.

Tail Bobs

Tail bobbing is a quick temporary way to identify cattle. Simply snip off the switch (below base of tail) with a knife or shears. Sick cows, cows in heat, culled

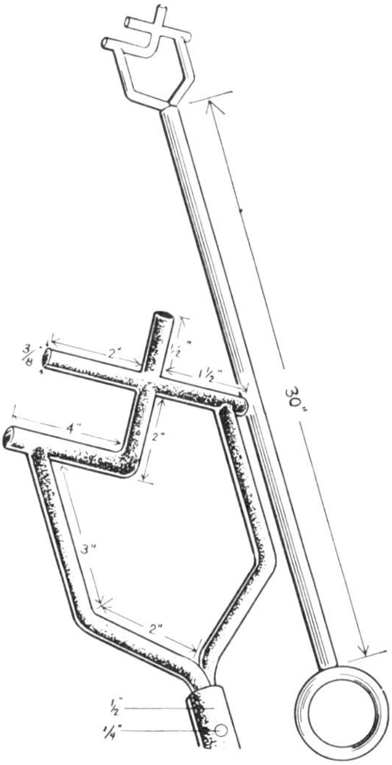

Fig. 4.11—Branding iron made from old hay rake tooth. This diagram was taken from University of California Agricultural Extension Service publication *How to Identify Livestock*. In addition to information on branding, this publication by Reuben Albaugh, Extension Animal Husbandman, contains information on earmarks, hair brands, horn brands, eartags, and tailbobs.

Fig. 4.12—A desirable way to earnotch a calf. Note how a small board is placed on the operator's knee so the ear can lie flat against it.

animals and so on can be identified from the rest of the herd by this procedure.

6. Reading Brands*

The use of brands as a mark of identification dates back some 4,000 years. Inscriptions and picture writing on the walls of ancient Egyptian tombs indicate that cattle were branded as early as 2000 B.C.

History also shows that use of the branding iron has not always been restricted to the hides of livestock. Slaves, fugitives and criminals were often stamped for life with the indelible mark of the brand. The practice of branding human beings was followed in England until as late as 1822.

The American custom of cattle branding was adopted

*Much of the information on this activity is from the pamphlet "History of Cattle Brands," courtesy of Humble Oil & Refining Company, Denver, Colorado.

from Mexico. The Mexican Dons marked their huge herds with their family coat-of-arms and, as the cattle industry moved northward into Texas, this method of indicating ownership gradually became accepted by American ranchers.

Today there are hundreds of thousands of cattle brands registered in the United States. When a rancher decides upon the type of brand he wants to use, the legal procedure is to register his mark. State laws designate a brand inspector or similar official who is responsible for assigning and recording brands. In some states registration must be made with the county clerk in each county or counties where the rancher expects to operate. This makes it possible for men in different parts of the country to use the same brand, although the reading of the brand may differ from one locality to another.

Read from Left to Right

Brands are read from left to right, from the top down, or from outside inside. A definite method of identifying characters has been established. If a letter or symbol is made backwards from its normal position, it's read as a reverse F, or whatever other letter it might be. A letter partially over on its face or back is said to be "tumbling." If a letter lies horizontally on its face or back, it is called "lazy." Letters with a curving flare at the top and rounded angles are called "running." Adding a dash to the left and one to the right at the top, you have a "flying" letter. Add legs and it becomes a "walking" letter. A letter placed so that the bottom touches the inside of a curve is said to be "rocking." In this particular instance the brand would probably be called the "anchor." Curves not attached to letters are known as "quarter circles" or "half circles," depending on the arc. Letters or symbols formed together are called "connected"

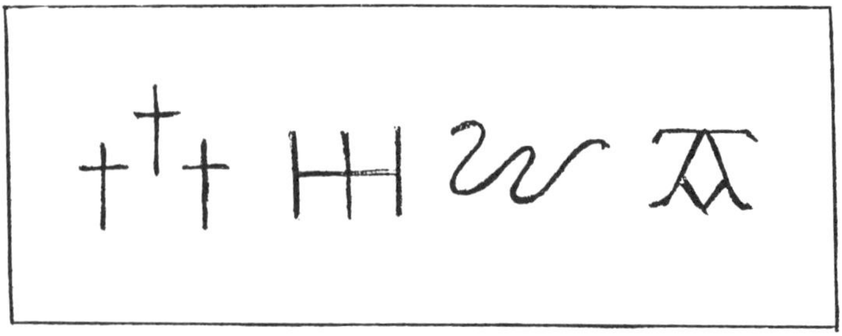

Fig. 4.13—Four historic brands in America. Reading from left to right: Three crosses, the brand of Hernando Cortez the Spanish conqueror and the first brand in America; double H the brand of baron Henry Miller of the well-known Miller and Lux estate; the running W of the famous King ranch in Texas; and the Spanish TA which is one of the oldest brands in continuous use in California. The owner is Julius Trescony of San Lucas, Monterey County.

, except when one is below the other; then the lower symbol is said to be "swinging." In registering brands, owners sometimes omit the "connected" or "swinging." Thus,  might be read simply Diamond J rather than Diamond Swinging J.

Memorize Character Brands

Besides the traditional letter and figure brands, there are some marks known as "character brands." For instance, this  is read as the turtle brand. A few other commonly used picture brands are: the pitchfork , the rocking chair , the key , the spade 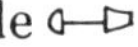, the ladder , and the signpost .

The reading of picture brands depends upon the owner's interpretation, and it takes an expert to identify some of the complex marks.

Try Angle Brands

ARS veterinary scientist R. Keith Farrell, Pullman, Washington, designed a system based on angles which

Symbols Commonly Used in Brands

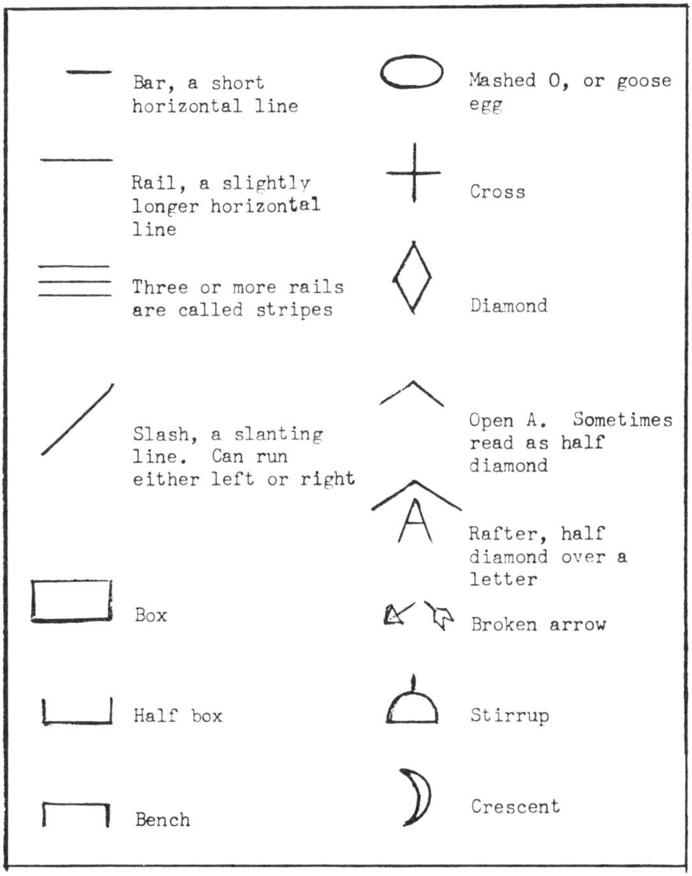

Fig. 4.14—Learn the above symbols. Most brands can be read if you are familiar with the common symbols.

See If You Can Read These Brands

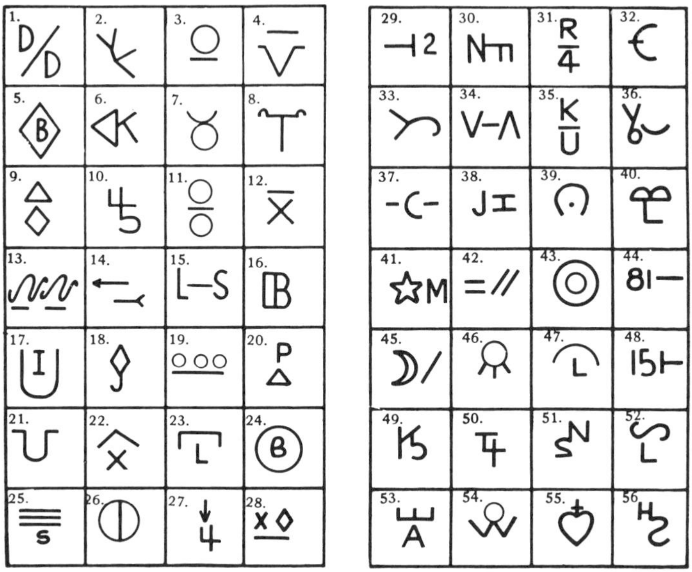

Courtesy, Humble Oil and Refining Co.

Fig. 4.15—Can you read the above brands? See Appendix B for answers.

is unalterable and easily read from a distance. (See Figure 4.16.)

The system is based on two concentric squares. In the basic square, numbering starts in the upper left corner with two, and continues clockwise with four, six, and eight. The odd numbers are obtained by rotating the basic square an eighth turn to the right so the corners fill in the gaps between the even numbers. Numbering begins at the top with three and runs clockwise in five, seven, and nine.

Various combinations of angles, with vertical and horizontal parallel lines for one and zero, respectively, are used to form symbols. Position of the angles rela-

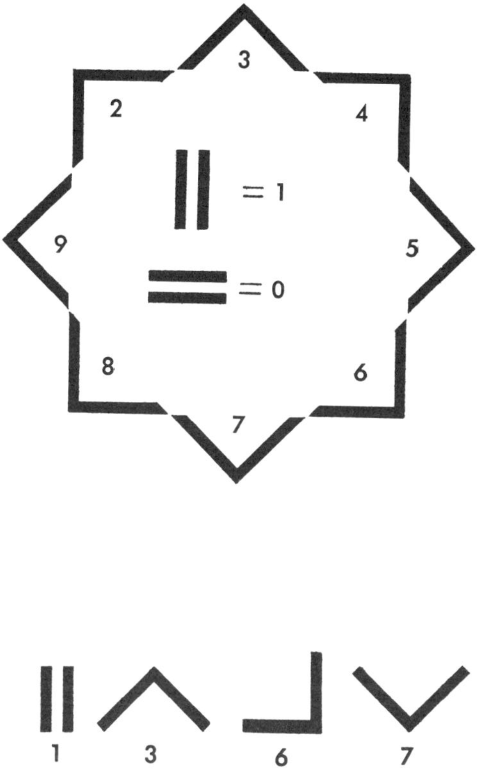

Fig. 4.16—Top: Master "clock face," formed by two concentric squares, shows the angles, and the numbers they represent. Bottom: This brand, number 1367, may be read by matching angles against the clock face.

tive to the concentric squares indicates how the symbols are read.

Animals were marked with this system by tattoo, freeze brand, and laser.

In recognition trials, the angles were more easily read than numbers, even though people are familiar with numbers. For example, at 40 feet only 59 per cent of the numbers were recognized, compared with 90 percent of the angle symbols.

7. Creep Feeding Calves

Creep Feeding Desirable—In this practice, young calves are allowed access to extra feed while they are still nursing. Plans for constructing creep feeders are found in the chapter on Beef Equipment. This practice enables calves to do well even though their dams are poor milkers, and indirectly helps the cow when feed conditions are unfavorable because the calves draw less heavily upon their mothers. Generally, calves begin to eat grain at four to five weeks of age.

Provide suitable rations for the calves in the creep. Farmer's Bulletin 1916 recommends the following rations:

(1) Young Calves:

yellow corn or ground barley	30	parts by weight
ground oats	25	" " "
wheat bran	25	" " "
linseed meal	8	" " "
dried skim milk	9	" " "
salt	2	" " "
steamed bone meal	2	" " "

(2) After 3 months of age:

cracked corn	4	" " "
ground oats	3	" " "
wheat bran	2	" " "
linseed meal	1	" " "

Well-bred calves when properly creep fed attain a weight of 700 to 800 pounds and desirable market finish at ten to twelve months of age.

Creep Feeders Admit Calves Only—Calves can enter creep feeders—without interference from full-grown cattle—to get grain as soon as they will take it. Adapt basic idea to local needs.

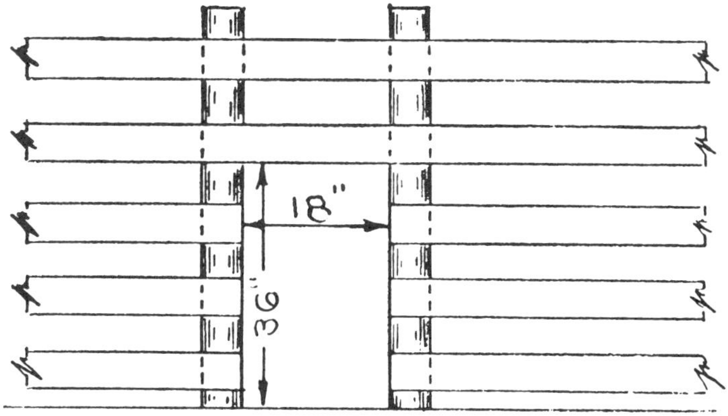

Fig. 4.17—Suggested dimensions for a desirable creep.

Creep Feeding Not Adaptable to Rough Areas—In order to be practical, creep feeding must be used where the herd comes regularly to a central point for water, salt, or shade. On rough, hilly areas or where cattle graze over vast areas, the method is generally impractical.

8. Providing Proper Minerals

Minerals Essential—Minerals are necessary for the health of animals and even for life itself. The amount and kind of minerals will depend to a large extent on the kind of ration the animals are receiving and the area on which they are grazing. It is just as important to avoid the extra expense and trouble of feeding minerals as it is to supply them when they are necessary.

Provide Salt "Free-Choice"—Salt is the mineral

most needed by beef cattle and "free-choice" is the logical way of feeding it. Beef breeding cows, especially when nursing calves, eat more salt than fattening steers. When cattle are on pasture, especially on succulent pasture, early summer pasture, which is rich in potassium, they consume more salt. Therefore, special care should be exercised to provide plenty of salt at these times. Many livestock men prefer half-ground rock salt.

SUGGESTED CONCENTRATE MIXES

For feeding with good afalfa hay in drylot

Ground or rolled grain	75 lb.
Dried molasses beet pulp	20 lb.
Cottonseed meal (with excellent alfalfa hay, cottonseed meal may be dropped)	5 lb.

For feeding with oat or other grain hay in drylot

Ground or rolled grain	70 lb.
Dried molasses beet pulp	15 lb.
Cottonseed meal	10 lb.
Alfalfa meal	5 lb.

For feeding with irrigated or spring dryland pasture

Ground or rolled grain	80 lb.
Dried molasses beet pulp	20 lb.

Provide Ample Calcium and Phosphorus—Requirments of calcium and phosphorus will be met by most otherwise well-balanced rations. Rations with heavy concentrate and minimum amounts of non-legume hay will need a calcium supplement. Ground limestone or ground oyster shell is a good source of calcium. Rations consisting of alfalfa hay, molasses, and beet pulp with little or no grain will need phosphorus supplements. Steamed bone meal is a good source of phophorus. These minerals may be fed free-choice, mixed with salt or combined with other feeds.

Provide Other Necessary Minerals—Occasionally

smaller amounts of other important minerals must be included in the diet if proper growth is to be made. In some areas, like those in Northern California or in the Midwest, iodine may be lacking in the drinking water. If such is the case, it may cause goiters in cattle, and calves will be born weak or dead. Iodine is best supplied by frequently mixing fresh potassium iodide with ground rock salt.

Some areas, like those in Florida, may lack cobalt or other trace elements that are necessary for the life and health of beef animals. Local inquiry to the state extension service or U.S.D.A. should be made concerning trace elements and how to supply them, as no advantage is gained in adding these minerals if they are unnecessary.

9. Properly Handling the Calves

This is especially important in a registered herd where animals are raised for show and for sale as individual breeding animals.

Practice "Gentling" the Calves—Be quiet around your calves and get them gentle by scratching or rubbing them. The use of a curry comb will further gentle them.

Practice Training the Calves—Teach the calf at an early age to stand and lead. Halter breaking, brushing, and clipping go hand in hand with feeding. Buyers are always impressed by a well-trained animal. Of course, this practice is unnecessary in commercial herds where all individuals will be sold as feeder or fat cattle.

Properly Train Horns—On horned breeds, the appearance of the animal is improved if the horns are trained to grow downward and slightly forward in a gentle curve. Training should be done with light weights

Fig. 4.18—Horn weights are used to shape horns. They should be removed after the curve becomes apparent in the shape of the horn.

when the horns are small or about 4½ to 6 inches, but solidly attached to the skull. One-half pound weights are about right with which to start. Care should be used so that too sharp a downward bend is not made.

Separate the Males from the Females—Bull calves should be separated from the heifers at about six months of age. In commercial herds, separation is unnecessary although many producers find it a desirable practice to separate steers and heifers after weaning.

Trim the Feet Early—Ill shaped feet affect the ap-

Fig. 4.19—Trimming feet with chisel. The rope used to secure the foot should always be tied to the pastern—never to the shank.

pearance and sale value and the way an animal walks. Trimming the feet will be necessary as the toes begin to grow. The most satisfactory way is to utilize a nipper, wood chisel, and wood rasp while the animal is securely held in foot trimming stocks. This operation is rarely necessary when cattle are on the range.

If only a small amount of trimming is necessary or perhaps only one foot is to be trimmed, it can be done easily, using longhandled pruning shears, while the

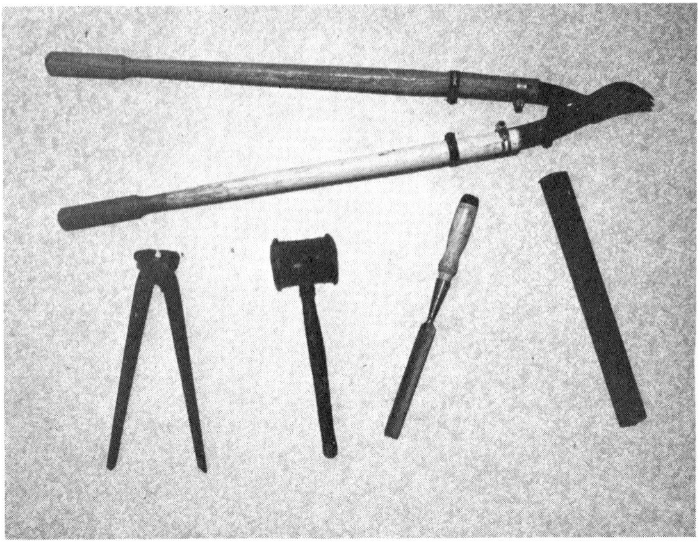

Fig. 4.20—Tools needed to trim feet: Pruning shears, hoof nippers, chisel, hammer or mallet, and rasp.

animal is standing on the ground. In order to do so, place the animal in a chute or alley so it cannot turn around, or else halter and tie it. While the animal is standing, any unwanted long horny growth can be clipped off. However, if extensive trimming is needed, then the animal must be restrained enough so the hoof can be cleaned, trimmed, and rasped level. See Fig. 4.19 which shows how the hoof is secured when the animal is restrained in a stock.

Inspect Feet of Mature Cattle Regularly—This is especially critical with old cattle and on irrigated pastures or during the winter on wet ground. Range cattle on rough ground generally wear off the hooves rapidly enough so that little trimming is necessary. However, older cattle and very gentle or inactive cattle frequently need to have their feet trimmed.

Provide Exercise and Sunshine—Sunshine provides Vitamin D. Calves should not be kept in a barn continually. Provide an exercise lot or grass pasture.

Check Snow Burn—Glare or snow-reflected sunshine in range country can blind whiteface calves or blister light colored teats and udders of nursing cows.

To prevent eye trouble paint or stamp a strip about one inch wide under each eye using nyanzol dye. Wyoming County agent, Nels Dahlquist recommends preparing the compound by mixing a measured quantity of powdered nyanzol in boiling water. Dissolve an equal amount of gum arabic (acacia gum) in cold water, mix the two together and place in a tight container. Before using, add half a cup of hydrogen peroxide. Keep above 40°F. or it will crystallize. The stain lasts as long as the hair lasts on a calf.

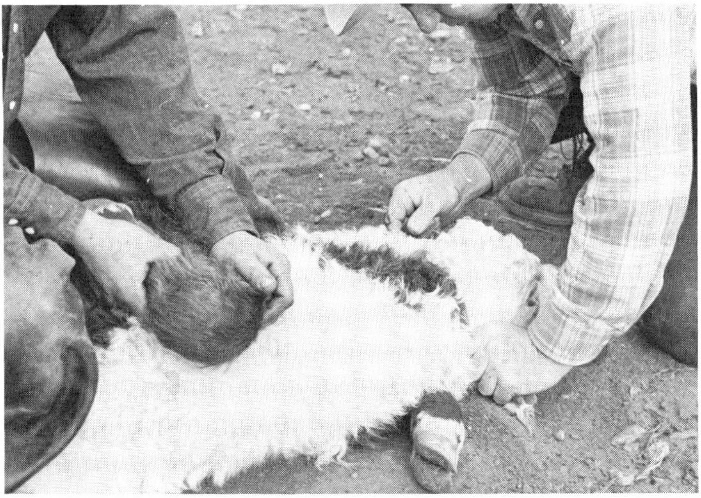

Fig. 4.21—Learn to properly restrain animals. Notice how the calf is "eared down" and the knee is bent and held by the assistant.

10. Dehorning and Castrating Calves

According to Farmer's Bulletin 1600, U.S.D.A., the following are desirable practices:

Dehorning, castrating, branding, marking, and vaccinating may be performed most satisfactorily and with greatest speed when a well-constructed chute is provided for holding the animal. Plans for this equipment are included in Chapter VI or in the *Handbook of Livestock Equipment,* published by The Interstate Printers & Publishers, Inc., Danville, Illinois.

A. Castration

Properly Castrate Calves—Castration improves the beef animal from a beef standpoint causing greater development of the regions from which valuable cuts are taken. Castration also assists in management of the herd.

One method is to grasp the scrotum, stretch it out tightly and cut off the lower one-third of the sac. One testicle should be removed at a time. To accomplish this, press the testicle out of the scrotum and cut it off allowing three or four inches of cord to remain on the testicle. Sever the cord by scraping it with a knife. Castration should not be performed by an inexperienced person.

Using the Bloodless Method—This calls for a special pinch or clamps which crush each cord separately an inch or two above the testicle.

Use the Elastrator—The younger the calf the better. Best results are obtained on calves from a few days old to a month old. At this age the skin is soft and the testicles will drop off more quickly. However, the Elastrator may be used on calves three or four months of age, if desired. The older the calf, the more the ring must be stretched and the longer it takes the purse to drop off.

Follow the manufacturer's directions for applying the ring to the scrotum. Be certain both testicles are down and through the ring before releasing. Bloodless methods are desirable if there is danger of screwworm infestation.

Protect Against Tetanus—If the elastrator band is not put on properly or the testicles are too large, as on older calves, there is danger of tetanus infection. This disease is generally fatal, so every precaution should be

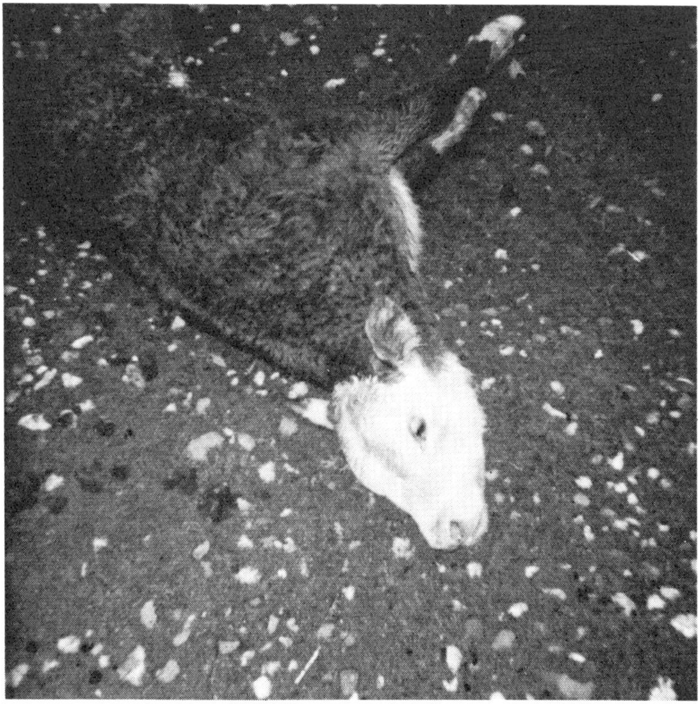

Fig. 4.22—Calf suffering from tetanus infection. Any sudden movement or jolting causes the animal to go into spasms. Vaccination at this time is useless.

taken to guard against its happening. Tetanus results from a deep wound where there is no oxygen, so conditions are set up where the germ can develop. Therefore, at the slightest indication of trouble, such as lack of quick healing, a deep wound, or swollen testicles, the calves should be immediately vaccinated for tetanus. Very valuable calves should be vaccinated for tetanus along with the castration operation. Treatment with antitoxin after the animals get tetanus is impractical and seldom successful. Consult a veterinarian if there is any question on procedure.

Select the Proper Time—Bull calves may be castrated from a few weeks up to four months of age. They should be castrated before they are weaned and generally as young as possible if consistent with herd management.

Cool weather is desirable to prevent fly strike, but the operation should not be done in severe weather.

Prevent Fly Strike—Fly strike or screwworm infestation should be guarded against. Pine tar oil smeared on the wound is generally effective. However, U.S.D.A. smear 6Z is considered at present most effective.

Spaying Heifers Not a Desirable Practice—Spaying is the removal of ovaries in females. Veterinarians are employed to do the work. The added expense and risk seldom justify the operation.

B. Dehorning

Horned feeder cattle require more space at the feed trough. Dehorned cattle produce carcasses with fewer bruises and better quality hides than horned cattle handled under the same conditions.

Dehorn at Proper Age—The younger cattle are when

they are dehorned, the better beef animals they will make. Dehorn during cool weather, if possible, to avoid fly strike danger.

Use Proper Method of Dehorning—If calves are from three to ten days of age, the best method is the application of caustic (or any commercial dehorning paste) to the horn "buttons" so as to prevent growth. The hair should be clipped away from the button and the area rubbed well with the caustic stick. Petroleum jelly can be applied to the surrounding area so as to

Fig. 4.23—Cattle that have been dehorned with dehorning paste as calves result in smooth heads such as above.

Fig. 4.24—Dehorning calf with caustic paste.

prevent skin burn. Calves so treated should be protected from rain for a few days. Range producers often have a herd rider constantly with the cattle, who dehorn all calves a few days after they are born. Saws or mechanical dehorners are used to dehorn mature cattle.

If saws are used, the cut should be made from the bottom of the horn upwards, and cutting toward the top or poll of the animal's head. A "squeeze" that securely holds the animal's head is almost exclusively used to hold the animals. This device is illustrated in the section on beef equipment.

Prevent Excessive Bleeding—On mature animals, profuse bleeding may take place. Apparently considerable bleeding can take place with little or no damage.

Excessive bleeding can be stopped with tannic acid or commercial clotting powders.

11. Properly Vaccinating Calves

Death losses can be minimized by proper immunization against disease. Prevention of diseases is most satisfactory by vaccination when carried on as part of the program of raising and producing calves and feeders.

Select the Proper Age—Depending on the disease, the age will vary. However, if possible, calves should be vaccinated as young as possible and before they go

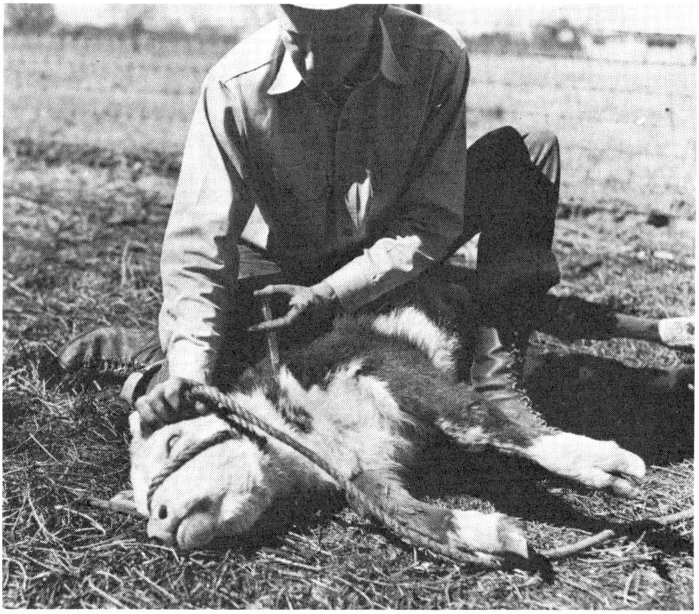

Fig. 4.25—Vaccinating calf while it is still young enough to be handled easily.

on summer range. It is well to do any such operation before weaning.

Vaccinate Against Blackleg—Blackleg immunization is almost a universal practice. Protection through immunization should be done as early as possible in accordance with directions supplied by manufacturer of vaccine.

Guard Against Anthrax—In anthrax-infected areas, routine vaccination is necessary. Veterinary service should be employed immediately in all anthrax outbreaks. See chapter on disease.

Vaccinate Heifers for Bang's Disease—Bang's disease is not as great a problem in beef cattle as with dairy herds. However, most progressive breeders follow a definite pattern of control. Heifers should be vaccinated with an approved vaccine at six to eight months of age. This should accompany a regular program of blood testing the herd. Some states now have a system of calfhood vaccination similar to the T.B. control program. Sometimes this is without cost to the beef producer.

Cooperate with T.B. Inspection Program—Federal inspection and control of tuberculosis has received national acclaim. Continued cooperation with authorities and watchfulness should be practiced.

12. Providing Satisfactory Pastures

Good Pasture Essential—Pasture, either irrigated or not, provides the key to economical beef production. Rarely is a profit made with farm beef cattle without pasture supplying the bulk of the feed until cattle reach the feed lot.

Irrigated pastures under good management provide:

a. Conservation and improvement of soil resources.

b. Cheaper, more nutritious feed for livestock.

c. Low harvest costs because livestock harvest their own feed.

d. Some saving in protein feed supplements.

e. Better gains in young stock.

f. Higher production in a breeding herd.

g. More economical gains in market animals.

h. A farm labor load more evenly distributed over the year.

Unirrigated native pastures have many of these same advantages. Cattle raised on adequate pastures seldom have as serious a disease or parasite problem as those raised in confinement. Therefore, approved practices on pastures should be followed.

Fig. 4.26—Cattle grazing irrigated pasture. Plants should be partially mature before cattle are turned in to graze.

Properly Utilize Pasture—A common saying is to "keep the feed ahead of the cattle." This means the number of head feeding on pasture should balance the available feed supply with some reserve so as not to damage the pasture. However, there is no advantage in keeping too few head on a given area, as the feed may grow too rank and coarse so that cattle dislike to eat it. Furthermore, grass that is not consumed one season generally is spoiled by rain, thereby becoming a loss to the producer.

Do Not Overgraze Pastures—While it is important to utilize feed to the capacity of the pasture, much damage can result by overgrazing. Most cattlemen feel that they can prevent overgrazing by one of two methods.

a. Rotation of pastures—Move cattle from field to field so as to utilize feed, but remove them before the pasture is damaged. This method requires adequate fencing of fields.

b. Sensible continuous grazing—Leave cattle on the area continually, but regulate numbers; place salt and water strategically and supply supplement when necessary so as to properly control grazing.

Regulate the Distribution of Minerals and Water—By developing water holes, digging wells, or otherwise providing water, you can have the cattle graze over a larger area and utilize more feed than if they must congregate in one locality to drink. Salt and minerals, as well as protein supplements, can be properly distributed over the range so as to make the cattle graze evenly over the pasture. Proper distribution of these items will also prevent useless walking on the part of the cattle so that less feed will be required for maintenance.

Provide Proper Fencing—Proper fencing insures the proper use of pastures, lessens the amount of labor required, and prevents animals from injuring themselves. Three- or four-strand barbed wire fences are adequate for control on the range, but for smaller areas near the farmstead, woven wire should be used. Fences around any areas into which the cattle are to be crowded or worked should be of heavy board construction.

Adjust Cattle Numbers to Feed Supply—Following are some practical methods of adjusting to seasonal and annual variations in feed supply:

a. Provide for a proper balance between summer and winter feed supply. This is essential when cattle are maintained on a single ranch unit as when winter feed and summer feed are supplied by separate units.

b. Stock the ranges moderately so that ample forage is available in average years for maximum gains, and excess forage waste through non-use is avoided.

c. Build up a reserve of hay and other feeds in good years and in times of low prices to meet drought and other extreme conditions.

d. Maintain some reserve of fat on breeding cattle so that weight losses can be withstood in an emergency without interfering with further production.

13. Properly Weaning Calves

Calves dropped early in the spring under favorable conditions will attain sufficient size and weight by summer or fall to be sold either as fat calves, stockers, or feeder calves, or they may be weaned when six to nine months old and kept within the herd. Early spring calves should never be permitted to follow their

Fig. 4.27—A well-built fence suitable for handling cattle.

mothers during the winter months. They may be kept quiet when weaned by placing them in a herd of dry cows. Following are several suggestions for weaning calves:

a. Wean purebred calves later.

Fig. 4.28—Light aluminum gates have become very popular. Gates should not sag, drag the ground or have sharp edges which might injure livestock.

In order to insure maximum growth, purebred calves nursing their mothers should not be weaned until eight to nine months old.

b. Weaning creep-fed calves is a simple matter.

This is true because the calves are advanced enough in their feeding period so that they do not miss their mothers. Furthermore, if the same creep-fed ration is continued, calves experience little set-back from weaning.

Fig. 4.29—A well-grown calf about ready to be weaned.

14. Producing Feeder Cattle

Feeder Cattle Essential—With the exception of purebred breeders who are producing registered cattle for breeding stock only, a great bulk of the cattle raisers are producing primarily feeder cattle. Strictly speaking, feeder cattle are cattle almost mature and in good enough flesh to be economically fed to fatten for slaughter.

Several fundamental rules should be followed if feeds are to be marketed most efficiently through the production of feeder cattle.

a. Adopt a consistent, systematic breeding and culling program as has been outlined in previous chapters.

b. Secure through breeding, management, feeding, and disease control a high percentage calf crop.

c. Promote through feeding and management normal and continuous growth of young cattle to secure efficient feed utilization and a desirable product.

d. Adopt production and selling policies that will yield the highest return from the annual feed supply. This rule will be discussed in a later chapter.

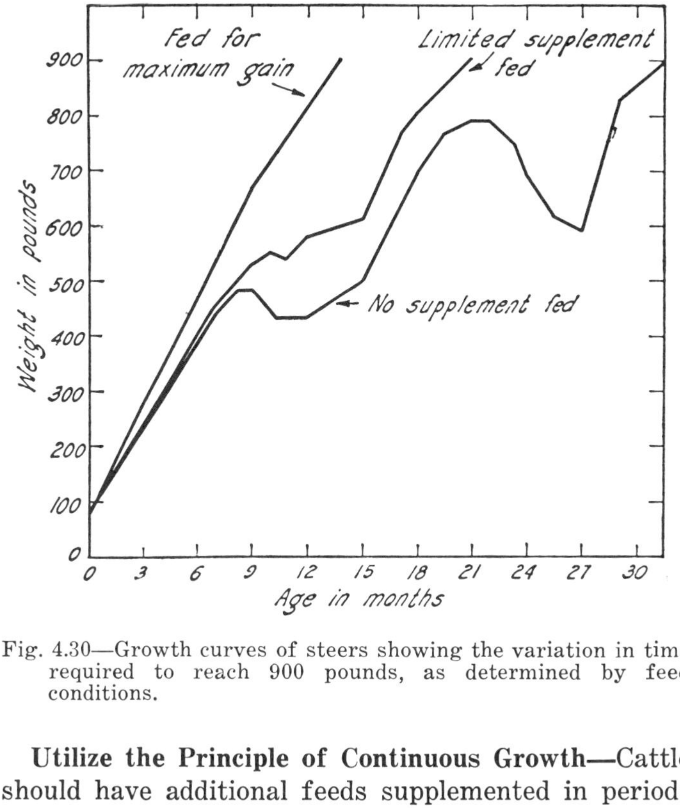

Fig. 4.30—Growth curves of steers showing the variation in time required to reach 900 pounds, as determined by feed conditions.

Utilize the Principle of Continuous Growth—Cattle should have additional feeds supplemented in periods of poor grazing so that they will not lose the gain they have made under good feeding conditions. This is especially important in young animals as they generally make the most economical gains in weight. Much food is required for maintenance; therefore, if cattle are kept growing continuously even throughout the winter months, the total amount of feed required per pound of beef will be less because they will reach a marketable age much sooner.

Supplement Calves and Yearlings on the Range—A gain of about one pound per day for calves that will

Fig. 4.31—Providing supplemental feed for cattle on the range.

be sold as feeders at the end of the next grass season is a sound objective. If steer calves are to be finished as long yearlings by feeding on grass, $1\frac{1}{4}$ to $1\frac{1}{2}$ pounds daily gain during the winter are desirable. One method of reaching this goal on dry range is to supplement 1 to $1\frac{1}{2}$ pounds daily of cottonseed cake or equivalent protein feeds.

Hay must be supplied in addition in areas where dry winter forage is not available. Steer calves so fed should weigh 850 to 900 pounds at the end of the next grass season.

Properly Winter Weaners and Yearlings on Harvested Roughages—No class of cattle responds more profitably to liberal feeding than calves during their first winter.

a. Provide sanitary mangers and feed bunks.
b. Provide well-drained corrals and shelters against winds and storms.
c. Utilize one of the following systems:
   (1) Supply all the high quality legume hay the calves will eat. Calves will eat 10 to 15 pounds daily depending on size.
   (2) Supply $\frac{3}{4}$ to 1 pound daily of cottonseed cake or equivalent with low-protein or non-legume hay.
   (3) Ten to 15 pounds of silage with all the legume hay the calves will eat will produce excellent winter gains without concentrates.

Calves so wintered should weigh 500-600 pounds in the spring, and 800-900 pounds at the end of the following pasture season.

Chapter V

FEEDING BEEF CATTLE

Feed Properly—Good feeding begins with proper management of the soil so it will produce maximum feed of good quality.

The feeding of cattle for beef production affords one of the most practical ways of disposing of grain and roughages produced on the farm. Diversification in agriculture and well-balanced use of farm labor is made possible by this industry.

Approximately 75 per cent of the fertilizing constituents of feeds fed to livestock is returned in the manure.

The feed produced by the land and the cattle on it are most productive when well-balanced rations are fed.

According to Farmer's Bulletin 1549, U.S.D.A., ability in buying and selling cattle and in combining feeds properly has a direct bearing on success in beef cattle feeding. Consequently, there are many approved practices in beef feeding which should be followed.

Activities Which Involve Approved Practices

1. Selecting desirable roughages.
2. Selecting desirable concentrates.
3. Providing satisfactory minerals and vitamins.
4. Providing sufficient water.

Courtesy Union Pacific Railroad

Fig. 5.0—These cattle will do well in the feedlot. They are uniform, without horns and well grown out.

5. Preparing feed properly.
6. Feeding the beef calf.
7. Feeding the breeding herd.
8. Feeding cattle for fattening.
9. Feeding beef calves for show.
10. Feeding and testing young bulls.
11. Providing for economical use of labor.
12. Providing comfort for animals.
13. Protecting cattle from stock poisoning plants.

1. Selecting Desirable Roughages

Coarse, bulky feeds called roughages, which oftentimes go to waste, can be converted into profits by feeding to cattle. It is important that land most suited for producing roughages yield large quantities of high quality feeds.

Provide Pastures—Good pastures are the foundation of profitable beef production because they furnish the cheapest possible source of nutrients and cattle harvest their own food. Properly managed, they will build soil fertility and help control erosion. In years of high-priced corn or other concentrates, pastures save feed and produce cheaper beef although rates of gain are slower on pasture than on full feed concentrates. Pastures fit in well with other crops in a rotation system.

Many pastures benefit from reseeding with new improved strains and from fertilization. Pastures should be cultured to produce the greatest number of pounds of beef per acre per year.

Pit Pastures

Pitting shortgrass ranges will increase forage growth and allow more animals to be run per acre. The pits made with an eccentric disk kill a portion of the range plants but the remaining plants yield more feed because of trapped moisture. The process should be repeated every 8 to 10 years.

Green Grass Best—Native pasture and grass produced on land too rough for extensive farming provide excellent feed during the grass-growing season. However, dry grass is lower in protein and supplemental protein feeds must be provided when dry grass is fed.

Practice Conservative Stocking—More cattle than the land is capable of carrying should never be allowed. Reserved grazing for use in dry periods is a good safety practice to follow. Native ranges should always be allowed to set seed. This is especially true for annual grasses. However, once annual plants produce seed, there is no advantage in not letting cattle clean up all the remaining forage by grazing.

Provide Permanent Pasture—Pastures planted to a mixture of perennial grasses and legumes provide an excellent source of economical nutrients. Furthermore, the growing plants are especially high in Vitamin A and proteins compared to dry pasture plants. A pasture livestock combination is the key to successful farming in many places. Some livestock producers obtain gains as high as two pounds per day per head, with 600 to 800 pounds of beef being produced on a single acre of grass pasture per year.

Land seeded to pasture does well in areas of summer rainfall or in the arid regions of the west and southwest where irrigation is practiced. Pastures capable of carrying over one-animal unit per acre are common in many locations.

Some common legumes used in pastures are ladino clover, red clover, trefoil, crimson clover, subterranean clover, white clover, and certain strains of alfalfa. A few of the many common grasses are tall fescue, orchard grass, Dallis grass, perennial rye, and bluegrass.

Provide a Mixture of Plants—Pastures with a mixture of grasses and legumes are less apt to produce bloat in cattle and provide a wide range of feeds over a longer period of time because certain plants do bettter than others at different seasons.

Properly Manage Pastures—Conservative continuous grazing or correct rotation of pastures will maintain the stand longer and produce more total pounds of beef. During the period of lush growth, the carrying capacity of irrigated pastures often exceeds two animal units per acre. However, between five and ten acres of these same pastures may be required to maintain one animal unit during the late fall and winter months. Therefore, it is a desirable practice to stock fields according to the right carrying capacity of the pasture. Proper fencing is a key factor in management of pastures.

Produce Grass-Fat Cattle—Good pasture, well managed, can produce grass-fat cattle as well as forming the reservoir as a source of feeder cattle. Many cattle each year are slaughtered without ever tasting feeds other than grass. Cattle produced on grass require very little labor in their production.

Properly Provide Hay—Choice hay should be provided for cattle at all times unless the animals are on good pasture. The extension service at Oregon State University estimates the average amount of hay consumed, in addition to pasture, annually by beef cattle in Oregon is one and one-fourth tons for each animal in the herd. Young animals consume less hay than mature animals. Where weather conditions are more severe or in high elevations, this amount will increase. If some concentrates or pasture are provided, a good general rule to follow is to feed one pound of good hay for each 100 pounds of live weight of animal.

Fig. 5.1—An electric fence is suitable to divide pastures for rotation of grazing.

Feed Good Quality Hay—Hay that is high in quality is leafy and cured from plants that are cut at an early stage of maturity. Stems should be fine and pliable. The color should be a bright green. Foreign matter and weeds should be absent and there should be no moldiness present. A fragrant odor characteristic of the crop from which it is made indicates hay that is high in quality.

Fig. 5.2—Many areas of the country have difficulty curing hay in order to yield high quality roughage. Here is one method used extensively in Canada, where rains are frequent. The hay is cut and baled as quickly as possible and left in these pyramid-shaped stacks to further cure as well as to be protected from sudden rains before they can be taken to shelter.

Purchase High Grade Hay—Because of the great bulk and expense involved in handling and transportation charges, it is a good practice to purchase the best grade hay in order to minimize these expenses. Oftentimes the best grade hay will cost no more per pound of nutrients than low grade hay, so the producer can feed the best hay at a lower cost than poor grade hay.

Feed Legume Hay If Possible—Hay made from legumes is high in protein, vitamins, and minerals, and is therefore preferred over other kinds of hay. However, hay made from grasses and other plants, if properly prepared, serves as an excellent roughage feed. The cost per unit of nutrients should always be considered when selecting hay to feed, regardless of the kind of hay, provided the hay crop has been properly cured.

Alfalfa—Alfalfa hay has no superior in feeding value for beef cattle. It is valuable for fattening cattle, breeding animals, and young stock. As it is high in calcium, phosphorus, protein, and vitamin A, it becomes especially desirable for breeding stock.

Other Legume Hays—While other legumes may not be quite as desirable as alfalfa, nevertheless they are excellent feeds for beef cattle and if high in quality, compare favorably with alfalfa. These include red clover, sweet clover, soybean, cowpea, vetch and oats, peanut hay, and others. Their feeding value may depend more on the manner in which the crop is cured than the kind of plant.

Brome alfalfa or clover-timothy mixtures are worth approximately 15-20 per cent less than straight legume hays.

Grass Hay—Hay made from the various grasses is an excellent beef feed although it is apt to be lacking in calcium and some protein; therefore, it should be supplemented. Equal parts of grass hay and legume hay make an excellent combination as the basic roughage feed for cattle. Oat hay, timothy, wheat, and barley are all good grass hays to be used for beef cattle.

Fig. 5.3—This pasture has been overgrazed. Note the undesirable plants appearing. Such pastures cannot supply adequate nutrients for cattle nor produce the maximum feed per acre.

Native western grass hays are excellent feeds for beef cattle and are in demand in these areas. Grass hay in some areas is considered equal to alfalfa from the standpoint of wintering cattle.

In the South, cane, Johnson grass, and Sudan grass are desirable hays. Sudan grass and associated grasses, such as sorghum, are excellent feeds for beef cattle under most conditions. Sudan grass is relatively drought resistant and, in many areas without summer rainfall or irrigation, yields excellent green pasture during the summer months when other grasses would dry up. If the crop is irrigated, it should not be pastured until the Sudan attains a height of 18 to 20 inches. Extreme caution should be exercised if growing conditions are not favorable, as prussic acid develops in the plant which is highly toxic to livestock. For this reason, Sudan is not recommended in areas where the growing season is short and frost is apt to be a factor during the pasture season.

Providing Silage—When forages are made into silage and fed to cattle, there will be little waste of plant foods provided the silage is of good quality. According to Farmer's Bulletin 578, silage has its greatest usefulness as the main feed so far as wintering rations for beef cattle are concerned. This applies particularly to stocker or feeder cattle and the breeding herd. Ordinarily silage contains 70 pounds water and 30 pounds dry matter per 100 pounds compared with hay, which is nearly 90 pounds dry matter. Therefore, a good rule to follow is to feed three pounds of silage per 100 pounds live weight of the animal.

Many crops can be made into silage, but where corn is grown, it is the outstanding silage crop.

Providing Root Crops—Root crops most generally grown for feeding livestock include mangels, rutabagas, turnips, and carrots. These do not differ greatly in composition, all having more water and less protein, fat, and carbohydrates than corn silage. They are chiefly valuable as a source of succulence and as appetizers, and are highly prized by breeders and exhibitors of purebred stock. Roots are sometimes fed whole, but when large, they should generally be cut. Ordinarily the limit in feeding is 3 to 5 pounds per 100 pounds of live weight, and the usual amount fed is generally not more than one-half this quantity. Excessive feeding may cause indigestion.

Haul Only Short Distances—Because of the large amount of water contained in root crops, it is generally not profitable to transport them more than a few miles.

Purchase Unharvested Crops—Many times root crops can be purchased at a ridiculously low price due to price failures or the crop not meeting market standards. Under these conditions, a source of cheap nutrients is available to the beef producer in that immediate area, and it should not be overlooked.

Utilize By-Products—Cattle are capable of converting large quantities of low cost unmarketable roughages into meat which can be sold. Therefore, it is a desirable practice to completely utilize all waste roughages on every farm.

Beet Tops—These consist of the cut off parts of the crown as well as the actual beet tops. Beet tops can be grazed by cattle directly in the field; they can be made into silage; or they can be cured by drying and then stored for future use.

Unfortunately for cattlemen, modern cultural practices of beef growers practically eliminate this once plentiful source of feed. The beets are now seldom topped in the field, as the entire root is harvested.

Price per ton of alfalfa hay	Per acre value of beet tops
$60.00	$197.00
65.00	213.50
70.00	230.00
75.00	245.50

The value of beet tops from a 20-ton-per-acre root crop at various prices of alfalfa hay. (The average top production from 20 tons of beet roots is three tons of dry-matter equivalent to 3.3 tons of alfalfa hay containing 10% moisture.)

Fig. 5.4—Even straw like that shown above can be profitably marketed through beef cattle.

Beef cattle make satisfactory gains on tops alone, but they are often supplemented with grains or legume hay. Beet tops are somewhat laxative. Animals should be under observation constantly in order to prevent death loss from choking on whole beets. Limestone should be provided for cattle grazing or consuming considerable amounts of this by-product. In western United States, this by-product is the basis for large cattle feeding operations.

Cottonseed Hulls—Cottonseed hulls are the roughage product of cottonseed oil manufacture. They have a very low protein content and should be fed only in connection with protein-rich feeds. As a roughage, the hulls have a lower feeding value than oat straw or corn stover, but are valuable where no other roughage is available. This product, however, is used extensively in the South, especially for steer feeding. Cottonseed products are lacking in vitamin A.

Crop Residues—Straws produced from small grains, such as oats, barley, wheat, and even corn stover, are a satisfactory roughage for beef cattle. They are especially valuable for wintering cattle because they are too low in nutrients to be used in fattening. However, if adequate grain is in the ration, they can be used fairly well to provide the roughage part in the fattening process. It must be remembered that straw is lacking in vitamin A, protein, and minerals.

Wintering trials at the Union Express Station in Oregon showed that cattle could be brought through the winter in reasonably good condition on grain straw and 1 to 2 pounds of oil cake daily. Straws from legumes and grasses contain more food value than grain straw.

The Montana Experiment Station found that 500 pounds of alfalfa and 1,500 pounds of straw wintered a cow without loss of weight.

Rice Straw—Any straw fed as a major portion of a growing ration for calves or yearlings leaves little if any energy available for growth after maintenance requirements are met. Work done at the University of California, Davis, however, indicates that rice straw is palatable and a satisfactory feed for pregnant cows provided protein, mineral, and vitamin requirements are met. Low phosphorus and energy intake by cows fed barley straw caused calving difficulties, but cows fed rice straw rations had no difficulty and produced calves within the normal weight range.

2. Selecting Desirable Concentrates

It is true that the bulk of the beef supply and a large portion of cattle slaughtered receive no feed other than grass. Nevertheless, concentrates form a most valuable asset to the self-feeding industry because of the following reasons:

First, they enable the beef producer to balance rations and thereby make more economical use of roughages. Secondly, concentrates are irreplaceable in conditioning or fattening cattle, thereby enabling the producer to raise the grade of his cattle and, in so doing, increase the selling price. Concentrates are low in fiber and high in total digestible nutrients.

Use Home-Grown Feeds—It is most economical to use home-grown feeds in supplying the needs of animals for concentrates or at least as the basis for the concentrate mixture. This is especially true if the roughage portion of the diet produced on the farm contains enough of the legumes so that the purchase of high-priced protein concentrate is unnecessary.

Provide Fattening Feeds—Fattening feeds are those concentrates that are high in carbohydrates and relatively low in protein.

Corn—This cereal grain is by far the most important fattening feed for beef cattle in the United States. It is very high in starches and fats, but low in protein and minerals, especially lime. Yellow corn is higher in vitamin A than white corn and therefore is often preferred. It is a very desirable concentrate for beef cattle and may be fed in a great variety of ways from "cattling down" to silage or pure shelled grain.

Barley—Barley contains slightly less total digestible nutrients than corn, but it is higher in protein. However, this depends upon the area in which it was grown. Barley is an excellent grain for fattening cattle and is superior to corn for breeding livestock. When

Fig. 5.5—Producing home-grown feeds is an essential practice for economical beef production.

fed to cattle, it is a desirable practice to coarsely grind or roll barley.

Wheat—Wheat contains almost twice as much total crude protein as corn and is higher in total digestible nutrients. For all practical purposes, wheat is a little more than equal to corn, pound for pound. However, since it is a heavy feed, it should be coarsely ground and fed with other bulky concentrates, such as ground ear corn or ground grain sorghum heads. It can be fed with safety when mixed with corn or sorghum on a 50-50 mixture basis.

Oats—Oats is somewhat lower in total digestible nutrients than corn or barley, and is, therefore, less valuable in a fattening ration. However, its bulkiness makes it especially desirable to include in the ration when cattle are starting on feed or for breeding cattle. Purebred breeders utilize oats as part of the ration in order to prevent digestive disorders in their cattle and keep the animal in proper health. Oats is often rolled or ground for feeding beef cattle.

Grain Sorghums—Grain sorghums are similar to corn in feeding value and wherever they grow well, are used as the main fattening feed. There are many varieties of sorghums including milo, kafir, and feterita, as well as hybrid varieties. All are about equal in feeding value. Sorghums are a desirable feed for beef cattle but should be ground before being fed.

Beet Pulp—Beet pulp dried or wet is an excellent feed for beef cattle and forms the basis for a fattening industry in the West. For fastest gains, some concentrates should be fed. Beet pulp is low in protein, phosphorus, and calcium. Therefore, some supplemental mineral feeding is necessary as well as a protein feed like alfalfa hay. There is no value in soaking dried beet pulp providing the animals have access to ample water.

Molasses—Beet or cane molasses is a valuable fattening feed if it can be obtained cheaply enough. It is unique in that it is almost pure carbohydrate. Care must be exercised in feeding molasses as too much has a decided laxative effect. Generally not more than 2 to 5 pounds per head daily is fed for this reason. One of the most desirable factors in feeding molasses is its use when sprayed on poorer quality roughages, thereby making them more palatable to livestock.

Select Most Economical Fattening Feeds—There are many feeds that can be utilized to feed and fatten beef cattle. Those selected for the ration should be based

RELATIVE VALUES OF FEEDS PER POUND FOR FATTENING BEEF CATTLE

Carbonaceous Concentrates	Value	Protein Supplements	Value	Roughages	Value
*Ground corn	100	*Soybean oil meal	100	*Alfalfa	100
Shelled corn (no pigs following)	85	Ground soybeans		Red clover	100
Ground barley	90	For calves	100	Soybean hay (10 per cent wastage)	90
Ground oats	86	For older cattle	130	Timothy (fair quality)	50
⅓ ground oats, ⅔ ground corn	95	Cottonseed meal (41 per cent)	100	Timothy (cut in early to full bloom, well-cured, high protein)	75 to 90
Ground wheat	100	Corn gluten meal	90	Oat straw	30
Ground rye as only grain	80 to 90	Linseed meal	130	Wheat straw	20
½ ground rye, ½ ground corn	100	Mixture ½ linseed meal and ½ other high protein supplement	130	Corn fodder	60
Hominy	100			Corn stover	35
Wheat bran (not more than ¼)	92			Corn silage, well-matured and well eared	50
Wheat middlings (not more than ¼)	100			Corn-stover silage	10
				Peavine silage	45
				Legume and grass silage (not wilted)	35 to 40
				Legume and grass silage (wilted before ensiling)	40 to 45

*Assigned a value of 100 From Cornell Univ. Ext. Bull. 581

on the cost per unit of nutrients available for food to beef cattle.

Select Proper Protein Feed—Protein feeds are most valuable in balancing rations, thereby enabling cattle to make the most efficient use out of the other feeds they consume. This is especially true of dry grass or concentrate feeds, such as corn, that are low in protein. Protein rich feeds added to such other feeds will make the total cost of production lower and gain in weight more rapid.

Note in the accompanying table how T.D.N. and D.P. decrease as crude fiber increases.

Cottonseed Meal—Cottonseed meal is used extensively for beef cattle as it is an excellent high protein concentrate. In the South it is often used as the only concentrate where it can be obtained cheaply. However, it is more often fed with a mixture of corn or other grains. The feed value is the same whether cake or meal, but under range or ranch conditions, cake is often

RELATIONSHIP OF TOTAL DIGESTIBLE NUTRIENTS AND DIGESTIBLE PROTEIN TO CRUDE FIBER

Crude fiber*	T.D.N.*	D.P.*
21.2	55	16.2
23.0	53	15.1
25.3	51	14.0
27.4	50	12.8
29.5	48	11.6
31.6	46	10.3
33.8	44	9.4

*90% D.M. basis

Courtesy, College of Agriculture and Environmental Science, Department of Animal Science, University of California at Davis.

preferred because it can be placed directly on the ground with little danger of blowing away. Cottonseed meal is lacking in vitamin A and calcium, therefore provision should be made in the ration for their addition.

Linseed Meal—This is an excellent protein feed for beef cattle. Purebred producers and those fitting cattle for show favor this concentrate because it produces rapid gains and gives a sleek, handsome appearance to the animals. While slightly lower in protein than cottonseed meal, it gives results fully equal or better than cottonseed meal under feeding conditions.

Soybean Meal—Soybean meal is a good protein feed for beef cattle. It is similar to cottonseed meal in feeding value although occasionally large amounts of it may be slightly laxative in their effect.

Other Protein Rich Concentrates—There are many other protein feeds, such as peanut meal, coconut meal, etc., valuable as feeds for beef cattle. Generally these should be fed only if they can be obtained cheaply enough and in conjunction with other feeds.

Wheat Bran—Wheat bran is not a high protein or high energy feed. However, it is greatly prized by cattle feeders and especially those fitting cattle for show because it is rich in minerals, bulky, and extremely palatable. For these reasons, it can be used to improve most otherwise well-balanced rations. Cattle feeders sometimes use wheat bran mixed with other concentrates to start cattle on feed.

Animal Proteins—Proteins from animal sources, such as meat scraps or tankage, are satisfactory feeds for cattle. These feeds are extremely high in protein and therefore take less to balance the ration. Cattle often refuse to eat them until they are camouflaged with other grains. In general, they are too highly priced to be extensively used in cattle rations.

For a complete analysis of the procedures to follow in determining the best ration, see the text *Approved Practices in Feeds and Feeding*, published by The Interstate Printers & Publishers, Inc., Danville, Illinois.

COST PER HUNDRED POUNDS OF PROTEIN IN FEEDS AT VARIOUS PRICES PER TON

Cost of feed per ton	Percent of protein in feeds										
	12	16	20	24	28	32	36	38	41	43	45
	Cents	*Cents*	*Cents*	*Cents*	*Cents*	*Cents*	*Cents*	*Cents*	*Cents*	*Cents*	*Cents*
$10.00	4.17	3.13	2.50	2.08	1.79	1.56	1.39	1.32	1.22	1.16	1.11
15.00	6.25	4.69	3.75	3.13	2.68	2.34	2.08	1.97	1.83	1.74	1.67
20.00	8.33	6.25	5.00	4.17	3.57	3.13	2.78	2.63	2.44	2.33	2.22
25.00	10.42	7.82	6.25	5.23	4.47	3.91	3.47	3.29	3.05	2.91	2.77
30.00	12.50	9.38	7.50	6.25	5.36	4.69	4.16	3.95	3.66	3.48	3.38
35.00	14.58	10.94	8.75	7.29	6.25	5.47	4.86	4.61	4.26	4.07	3.89
40.00	16.67	12.50	10.00	8.33	7.14	6.25	5.55	5.26	4.88	4.65	4.44
45.00	18.75	14.06	11.25	9.38	8.03	7.03	6.25	5.92	5.49	5.23	5.00
50.00	20.83	15.63	12.50	10.42	8.93	7.81	6.94	6.58	6.09	5.81	5.55
55.00	22.91	17.19	13.75	11.46	9.82	8.59	7.64	7.24	6.71	6.40	6.11
60.00	25.00	18.75	15.00	12.50	10.71	8.38	8.33	7.89	7.32	6.97	6.67
65.00	27.08	20.31	16.25	13.54	11.67	10.16	9.03	8.55	7.92	7.56	7.22
70.00	29.16	21.87	17.50	14.58	12.50	10.94	9.72	9.21	8.53	8.14	7.78
75.00	31.25	23.44	18.75	15.62	13.39	11.72	10.41	9.87	9.14	8.72	8.33
80.00	33.33	25.00	20.00	16.67	14.28	12.50	11.11	10.53	9.75	9.32	8.89
85.00	35.41	26.56	21.25	17.71	15.18	13.28	11.80	11.18	10.36	9.88	9.44

From U.S.D.A. Farmer's Bulletin 1179

AVERAGE NUTRIENT AND MINERAL CONTENT OF FEEDS

(all values except DE are decimals = pounds per pound of feed, or Kg per Kg of feed)

Feed	Dry Matter	Protein		Energy			Minerals	
		Total Crude	Digestible	Total Digestible Nutrients	DE Mcal/ Lb.	DE Mcal/ Kg	Calcium	Phosphorus
	1	2	3	4	5	6	7	8
Hay, Straw, and Similar Dry Roughages:								
Alfalfa hay, early bloom	.900	.166	.114	.510	1.026	2.262	.0112	.0021
Alfalfa hay, mid-bloom	.892	.152	.108	.520	1.035	2.281	.0120	.0020
Alfalfa meal, dehydrated (17% prot.)	.930	.179	.140	.580	1.154	2.543	.0133	.0024
Barley hay	.873	.078	.044	.500	.995	2.194	.0018	.0026
Barley straw	.882	.036	.004	.360	.723	1.595	.0030	.0008
Bermudagrass hay (Coastal)	.915	.087	.047	.400	.805	1.775	.0042	.0016
Bromegrass hay (smooth)	.928	.054	.018	.540	1.076	2.373	.0040	.0020
Canary grass hay (reed)	.913	.080	.042	.510	1.022	2.254	.0031	.0023
Clover hay (alsike)	.879	.129	.082	.530	1.055	2.326	.0115	.0022
Clover hay (crimson)	.874	.148	.103	.520	1.049	2.313	.0124	.0016
Clover hay (Ladino)	.912	.210	.132	.560	1.113	2.453	.0126	.0036
Clover hay (red)	.877	.131	.078	.520	1.035	2.281	.0141	.0019
Corncobs, ground	.904	.025	—	.420	.850	1.873	.0011	.0004
Corn fodder, sun-cured	.824	.073	.034	.540	1.071	2.362	.0059	.0025
Corn stover, sun-cured	.872	.051	.019	.510	1.029	2.268	.0043	.0008
Cotton bolls, sun-cured	.910	.091	.005	.380	.764	1.685	.0103	.0011
Cottonseed hulls	.903	.039	.002	.370	.741	1.633	.0014	.0009
Cowpea hay	.905	.166	.117	.570	1.140	2.514	.0121	.0029
Fescue hay (tall)	.885	.093	.053	.550	1.098	2.420	.0044	.0032
Gamagrass hay (Florida)	.923	.084	.042	.480	.960	2.116	.0057	.0029

(Continued)

AVERAGE NUTRIENT AND MINERAL CONTENT OF FEEDS (Continued)

Feed	Dry Matter	Protein: Total Crude	Protein: Digestible	Energy: Total Digestible Nutrients	Energy: DE Mcal/ Lb.	Energy: DE Mcal/ Kg	Minerals: Calcium	Minerals: Phosphorus
	1	2	3	4	5	6	7	8
Johnsongrass hay	.907	.070	.031	.510	1.016	2.239	.0073	.0028
Lespedeza hay (mid-bloom)	.930	.146	.098	.530	1.060	2.337	.0111	.0024
Meadow hay, intermountain	.929	.084	.027	.430	.855	1.884	.0053	.0016
Oat hay	.882	.081	.039	.540	1.076	2.372	.0023	.0021
Oat hulls	.930	.056	.020	.370	.744	1.640	.0016	.0019
Oat mill by-product	.920	.045	.023	.230	.460	1.014	0008	.0022
Oat straw	.901	.040	.013	.470	.937	2.066	.0030	.0009
Orchardgrass hay	.883	.086	.051	.500	1.007	2.219	.0040	.0033
Peanut shells, ground	.923	.068	.030	.070	.148	.326	.0023	.0006
Pea straw	.873	.066	.035	.500	.995	2.194	—	—
Prairie hay, mid-bloom	.910	.074	.037	.460	.910	2.006	.0031	.0019
Rye straw	.889	.027	—	.280	.551	1.215	.0025	.0009
Sorghum fodder (grain-type)	.903	.062	.023	.520	1.047	2.309	.0056	.0017
Sorghum stover (grain-type)	.851	.045	.015	.490	.970	2.138	.0034	.0009
Soybean hay	.892	.145	.090	.460	.928	2.045	.0115	.0020
Soybean straw	.876	.048	.015	.330	.665	1.467	.0139	.0005
Sudan grass hay	.889	.113	.049	.520	1.049	2.312	.0050	.0028
Timothy hay, mid-bloom	.884	.075	.041	.540	1.079	2.378	.0036	.0017
Trefoil, bird's-foot-hay	.912	.142	.098	.560	1.113	2.453	.0160	.0020
Vetch hay	.882	.176	.116	.550	1.094	2.411	.0120	.0030
Wheat hay	.859	.064	.029	.570	1.134	2.500	—	—
Wheat straw	.901	.032	.004	.430	.865	1.906	.0015	.0007
Protein Concentrates:								
Beans, kidney, dry	.890	.230	.154	.740	1.477	3.256	.0013	.0051
Blood meal	.910	.799	.567	.590	1.183	2.608	.0028	.0022

(Continued)

AVERAGE NUTRIENT AND MINERAL CONTENT OF FEEDS (Continued)

Feed	Dry Matter	Protein: Total Crude	Protein: Digestible	Energy: Total Digestible Nutrients	Energy: DE Mcal/Lb.	Energy: DE Mcal/Kg	Minerals: Calcium	Minerals: Phosphorus
	1	2	3	4	5	6	7	8
Brewers grains, dried	.920	.259	.191	.610	1.214	2.677	.0027	.0050
Buttermilk, condensed	.290	.107	.096	.270	.533	1.176	.0044	.0026
Buttermilk, dried	.930	.320	.288	.840	1.674	3.690	.0134	.0094
Casein, dried	.900	.818	.794	.850	1.692	3.730	.0061	.0099
Coconut meal, expeller	.930	.204	.165	.750	1.506	3.321	.0021	.0061
Coconut meal, solvent	.920	.213	.172	.680	1.362	3.002	.0017	.0061
Corn distillers grains, dried	.920	.271	.212	.770	1.546	3.408	.0009	.0037
Corn distillers solubles, dried	.930	.269	.210	.820	1.637	3.608	.0035	.0137
Corn germ meal, expeller	.930	.202	.148	.810	1.618	3.567	.0005	.0051
Corn germ meal, solvent	.930	.180	—	—	—	—	.0010	.0040
Corn gluten feed	.900	.253	.218	.740	1.476	3.254	.0046	.0077
Corn gluten meal	.910	.429	.357	.760	1.529	3.371	.0016	.0040
Cottonseed meal, 36% prot., expeller	.935	.396	.313	.860	1.720	3.792	.0019	.0102
Cottonseed meal, 41% prot., expeller	.940	.410	.332	.730	1.466	3.233	.0016	.0120
Cottonseed meal, 41% prot., solvent	.915	.410	.332	.690	1.373	3.026	.0016	.0120
Cottonseed meal, 50% prot., solvent	.925	.500	.404	.690	1.388	3.059	.0016	.0101
Cottonseed, whole ground	.927	.231	.146	.840	1.687	3.719	.0014	.0068
Crab meal	.930	.311	—	—	—	—	.1532	.0159
Distillers grains, dried	.916	.291	.231	.770	1.539	3.393	.0020	.0055
Eggs, chicken with shells	.341	.128	—	—	—	—	.0150	—
Feather meal, hydrolyzed	.910	.854	.602	.620	1.237	2.728	.0041	.0049
Fish meal, anchovy	.930	.660	.607	.680	1.358	2.994	.0450	.0285
Fish meal, menhaden	.920	.613	.496	.680	1.362	3.002	.0549	.0281
Fish solubles, condensed	.510	.314	.301	.440	.887	1.956	.0061	.0070
Fish solubles, dried	.920	.628	.603	.770	1.546	3.408	—	—
Flaxseed screenings, oil feed	.913	.241	.136	.540	1.077	2.375	.0044	.0063

(Continued)

AVERAGE NUTRIENT AND MINERAL CONTENT OF FEEDS (Continued)

Feed	Dry Matter	Protein: Total Crude	Protein: Digestible	Energy: Total Digestible Nutrients	Energy: DE Mcal/Lb.	Energy: DE Mcal/Kg	Minerals: Calcium	Minerals: Phosphorus
	1	2	3	4	5	6	7	8
Linseed meal, expeller	.910	.353	.310	.740	1.474	3.250	.0044	.0089
Linseed meal, solvent	.910	.351	.309	.690	1.383	3.049	.0040	.0083
Malt sprouts	.930	.262	.204	.640	1.283	2.829	.0022	.0073
Meat and bone meal	.940	.506	.461	.680	1.354	2.984	.1057	.0507
Meat meal tankage	.920	.598	.545	.730	1.453	3.204	.0594	.0317
Milk albumin, dried	.923	.473	.430	.460	.923	2.035	—	—
Milk, dried skim	.940	.335	.301	.810	1.617	3.564	.0126	.0103
Milk, dried whole	.937	.252	.244	1.170	2.343	5.165	.0089	.0068
Milk, fresh, skim	.096	.027	.026	.090	.179	.394	.0012	.0010
Milk, fresh, whole	.120	.031	.030	.160	.312	.688	—	—
Peanut oil meal, solvent	.920	.474	.427	.710	1.417	3.123	.0020	.0065
Peanuts, whole	.920	.191	.151	.990	1.987	4.381	—	.0030
Poultry by-product meal	.934	.554	.471	.740	1.480	3.263	.0300	.0170
Rapeseed oil meal, expeller	.936	.371	.318	.690	1.385	3.054	.0060	.0097
Rapeseed oil meal, solvent	.903	.394	.339	.620	1.246	2.747	.0040	.0090
Safflower oil meal, expeller	.910	.197	.145	.520	1.037	2.287	.0023	.0071
Safflower oil meal, solvent	.918	.214	.172	.500	1.010	2.226	.0034	.0084
Sesame oil meal, expeller	.930	.479	.383	.700	1.395	3.076	.0203	.0129
Soybean oil meal, solvent	.890	.458	.390	.720	1.442	3.178	.0032	.0067
Sunflower meal, expeller	.930	.410	.364	.650	1.302	2.870	.0043	.0104
Tomato pomace, dried	.920	.217	—	—	—	—	.0028	.0057
Torula yeast, dried	.930	.483	.439	.740	1.488	3.280	.0057	.0168
Wheat germ meal	.900	.262	.247	.860	1.710	3.770	.0007	.0104
Yeast, irradiated dried	.940	.481	.443	.700	1.410	3.108	—	.0128

*Taken from *Approved Practices in Feeds and Feeding*, The Interstate Printers & Publishers, Inc., Danville, Illinois.

NUTRIENT REQUIREMENTS OF BEEF CATTLE

(Nutrient Concentration in Ration Dry Matter)*

Body Weight kg (lb.)	Daily Average Gain kg (lb.)	Daily Dry Matter per Animal kg (lb.)	Total Protein (%)	Digestible Protein (%)	Energy		Ca (%)	P (%)	Carotene (mg/kg)	Vitamin A (Thousands IU/kg)
					ME (Mcal/kg)	TDN (%)				
Finishing Steer Calves										
150(331)	0.90(1.98)	3.5(7.7)	12.8	8.6	2.82	78	0.60	0.43	5.5	2.2
209(441)	1.00(2.20)	5.0(11.0)	12.2	8.1	2.67	74	0.46	0.34	5.5	2.2
300(661)	1.10(2.43)	7.1(15.7)	12.2	8.1	2.67	74	0.37	0.27	5.5	2.2
400(882)	1.10(2.43)	8.8(19.4)	11.1	7.1	2.67	74	0.28	0.23	5.5	2.2
450(992)	1.05(2.31)	9.4(20.7)	11.1	7.1	2.67	74	0.22	0.22	5.5	2.2
Finishing Yearling Steers										
250(551)	1.30(2.87)	7.2(15.9)	11.1	7.1	2.61	72	0.40	0.28	5.5	2.2
300(661)	1.30(2.87)	8.3(18.3)	11.1	7.1	2.61	72	0.35	0.25	5.5	2.2
400(882)	1.30(2.87)	10.3(22.7)	11.1	7.1	2.61	72	0.27	0.22	5.5	2.2
500(1,102)	1.20(2.65)	11.5(25.4)	11.1	7.1	2.61	72	0.23	0.22	5.5	2.2
Finishing Two-Year-Old Steers										
350(772)	1.40(3.09)	10.3(22.7)	11.1	7.1	2.56	71	0.29	0.22	5.5	2.2
400(882)	1.40(3.09)	11.3(24.9)	11.1	7.1	2.56	71	0.27	0.22	5.5	2.2
500(1,102)	1.20(2.65)	13.4(29.5)	11.1	7.1	2.56	71	0.22	0.22	5.5	2.2
550(1,213)	1.30(2.87)	13.7(30.2)	11.1	7.1	2.56	71	0.22	0.22	5.5	2.2

(Continued)

NUTRIENT REQUIREMENTS OF BEEF CATTLE (Continued)

Body Weight kg (lb.)	Daily Average Gain kg (lb.)	Daily Dry Matter per Animal kg (lb.)	Total Protein (%)	Digestible Protein (%)	Energy ME (Mcal/kg)	Energy TDN (%)	Ca (%)	P (%)	Carotene (mg/kg)	Vitamin A (Thousands IU/kg)
Finishing Heifer Calves										
150(331)	0.80(1.76)	3.5(7.7)	12.8	8.6	2.82	78	0.51	0.37	5.5	2.2
200(441)	0.90(1.98)	5.0(11.0)	12.2	8.1	2.67	74	0.42	0.30	5.5	2.2
300(661)	1.00(2.20)	7.3(16.1)	12.2	8.1	2.67	74	0.31	0.25	5.5	2.2
400(882)	0.95(2.09)	8.7(19.2)	11.1	7.1	2.67	74	0.26	0.22	5.5	2.2
Finishing Yearling Heifers										
250(551)	1.20(2.65)	7.6(16.8)	11.1	7.1	2.61	72	0.36	0.26	5.5	2.2
300(661)	1.20(2.65)	8.6(19.0)	11.1	7.1	2.61	72	0.31	0.23	5.5	2.2
400(882)	1.20(2.65)	10.7(23.6)	11.1	7.1	2.61	72	0.28	0.22	5.5	2.2
450(992)	1.10(2.43)	11.0(24.3)	11.1	7.1	2.61	72	0.22	0.22	5.5	2.2
Growing Steers										
150(331)	0.00(0.00)	2.7(6.0)	7.8	4.2	2.06	57	0.19	0.19	5.5	2.2
	0.25(0.55)	3.1(6.8)	11.1	7.1	2.28	63	0.26	0.23	5.5	2.2
	0.50(1.10)	3.2(7.1)	12.2	8.1	2.61	72	0.38	0.31	5.5	2.2
	0.75(1.65)	3.2(7.1)	13.3	9.0	2.82	78	0.53	0.41	5.5	2.2
200(441)	0.00(0.00)	3.3(7.3)	7.8	4.2	2.06	57	0.18	0.18	5.5	2.2
	0.25(0.55)	4.5(9.9)	10.0	6.1	2.06	57	0.18	0.18	5.5	2.2
	0.50(1.10)	4.9(10.8)	11.1	7.1	2.28	63	0.27	0.20	5.5	2.2
	0.75(1.65)	5.0(11.0)	11.1	7.1	2.50	69	0.36	0.28	5.5	2.2

(Continued)

NUTRIENT REQUIREMENTS OF BEEF CATTLE (Continued)

Body Weight kg (lb.)	Daily Average Gain kg (lb.)	Daily Dry Matter per Animal kg (lb.)	Total Protein (%)	Digestible Protein (%)	Energy		Ca (%)	P (%)	Carotene (mg/kg)	Vitamin A (Thousands IU/kg)
					ME (Mcal/kg)	TDN (%)				
Growing Steers (Continued)										
300(661)	0.00(0.00)	4.5(9.9)	7.8	4.2	2.06	57	0.18	0.18	5.5	2.2
	0.25(0.55)	6.1(13.4)	8.9	5.2	2.06	57	0.18	0.18	5.5	2.2
	0.50(1.10)	7.7(17.0)	10.0	6.1	2.06	57	0.18	0.18	5.5	2.2
	0.75(1.65)	8.0(17.6)	11.1	7.1	2.28	63	0.21	0.18	5.5	2.2
400(882)	0.00(0.00)	5.6(12.3)	7.8	4.2	2.06	57	0.18	0.18	5.5	2.2
	0.25(0.55)	7.7(17.0)	8.3	4.6	2.06	57	0.18	0.18	5.5	2.2
	0.50(1.10)	9.7(21.4)	8.9	5.2	2.06	57	0.18	0.18	5.5	2.2
	0.75(1.65)	9.9(21.8)	8.9	5.2	2.28	63	0.18	0.18	5.5	2.2
Growing Heifers										
150(331)	0.00(0.00)	2.7(6.0)	7.8	4.2	2.06	57	0.19	0.19	5.5	2.2
	0.25(0.55)	3.2(7.1)	11.1	7.1	2.28	63	0.25	0.22	5.5	2.2
	0.50(1.10)	3.2(7.1)	12.2	8.1	2.61	72	0.38	0.31	5.5	2.2
	0.75(1.65)	3.3(7.3)	13.3	9.0	2.82	78	0.52	0.39	5.5	2.2
200(441)	0.00(0.00)	3.3(7.3)	7.8	4.2	2.06	57	0.18	0.18	5.5	2.2
	0.25(0.55)	4.6(10.1)	10.0	6.1	2.06	57	0.18	0.18	5.5	2.2
	0.50(1.10)	5.0(11.0)	11.1	7.1	2.28	63	0.26	0.20	5.5	2.2
	0.75(1.65)	5.4(11.9)	11.1	7.1	2.50	69	0.33	0.26	5.5	2.2

*Taken from "Nutrient Requirements of Beef Cattle"; Publication 1137, Subcommittee on Beef Cattle, National Academy of Sciences, Washington, D.C.

Supply Nitrogen by Using Urea—Urea is generally a more economical source of nitrogen than natural sources of protein. According to the publication AS-344 Cooperative Extension Service, Purdue University, an increasing number of cattle feeders are relying on urea as the main protein supplement (or nitrogen source) in feedlot rations. Most combinations of farm-grown feeds for cattle are deficient in protein and must be supplemented. This additional protein is needed to balance a ration for maximum growth and performance. Bacteria in parts of a ruminant's stomach can utilize non-protein nitrogen if enough other nutrients (energy, vitamins, minerals) are available. However, feeding urea requires some special considerations in order to be an effective and economical supplement.

What Is Urea?—Urea is not a protein and contains no vitamins, minerals, or energy. It is a synthetic chemical compound manufactured by heating carbon dioxide, anhydrous ammonia, and water under pressure. Urea also occurs naturally in the urine of most animals.

CO_2	+	NH_3	+	H_2O	+	Heat → *UREA*
Carbon dioxide		Anhydrous ammonia		Water		

Urea nitrogen cannot be used directly by ruminants, but microorganisms in the rumen can use it directly to build bacterial protein. Non-ruminants (swine and poultry) cannot use urea even indirectly.

Billions of bacteria exist in the rumen, the first and largest compartment of the ruminant's stomach. If acceptable amounts of other nutrients are available, rumen bacteria can produce protein by combining urea nitrogen with carbon, hydrogen, and oxygen from starchy, high-energy feeds. The newly-formed protein then exists as part of the bacterial bodies.

UREA +	Energy + (carbohydrates; e.g., grain, molasses)	Rumen + Bacteria	Other Nutrients →	Body Proteins

Eventually, these bacteria pass into the fourth compartment of the ruminant stomach (abomasum). In the abomasum and intestines, the bacteria are digested and absorbed in the same manner as other protein.

As long as the rumen bacteria properly change urea nitrogen into bacterial protein, urea can be used efficiently as a protein supplement. If the bacteria cannot use the urea because of some limiting nutrient or because of an excess of urea, urea may then be detrimental to growth, performance, and overall health and may even cause death.

Replacement Value of Urea—Urea can usually reduce feed costs. However, before using urea the feeder should figure and compare the exact cost of different protein supplements.

One pound of 262 per cent urea has the same *crude protein equivalent* as 6.0 pounds of 44 per cent soybean meal. However, since oil meals contain energy as well as protein and since urea needs an additional energy source, direct price comparisons cannot be made.

Do not add urea to rations already containing sufficient protein. In addition, animals should be well started on feed before urea is added to the ration.

Sample Urea Supplements—The total protein in supplements may vary according to the amount of urea. On the chart atop page 185 are some supplements and recommended conditions for their use.

This ration is similar to Purdue Supplement "A" but contains 3.5% urea and may be fed at the rate of 1.0 to 2.0 pounds per day with moderate- to high-energy rations. If urea containing 45 per cent nitrogen is used,

Fig. 5.6—Cattle shown here are being fed urea in the form of range cubes. This is a non-protein nitrogen source for feed supplementation. The particular brand Kedlor releases ammonia slowly in the rumen to avoid toxicity problems.

PRICE COMPARISON OF UREA (262 PER CENT PROTEIN) PLUS CORN WITH SOYBEAN MEAL AS ENERGY AND PROTEIN SOURCES

Corn plus Urea	
1721 lb. corn × \$2.20/cwt. =	\$37.86
279 lb. urea × \$5.00/cwt. =	13.95
Total cost	\$51.81
Soybean Meal	
2000 lb. meal × \$4.00/cwt. =	\$80.00
Total cost	\$80.00
Difference in cost = \$28.19 per ton of supplement.	

the per cent of urea in the ration would be reduced to 3.3 per cent.

Two separate mineral boxes which contain (a) a mixture of 2 parts bonemeal or di-calcium phosphate and 1

32% SUPPLEMENT*

Ingredient	Per Cent
Urea (262% protein equiv.)	3.5
Soybean meal (44%)	42.5
Ground shelled corn	20.0
Cane molasses	12.0
Dehy. alfalfa meal (17%)	15.0
Bonemeal or dical. phosphate	5.2
Iodized salt	1.8
Total	100.0%
Micronutrients per 1000 Pounds	
Cobalt carbonate	2 gm
Zinc oxide (80% zinc)	625 gm
Vitamin A	10 mil.IU

*Plus feed additives in sufficient quantities to meet recommended daily amounts.

part iodized salt and (b) trace-mineralized salt should be available for the cattle.

The Missouri Station recommends feeding the following mineral mixture free choice: 40% iodized salt, 40% bonemeal, 20% limestone, plus 1 ounce of cobalt chloride and 0.5 pound of zinc oxide per 100 pounds of mineral mix. Trace-mineralized salt should also be offered in a separate box.

Cattle Performance on Urea Versus Natural Protein—If all the above conditions are satisfied, cattle should gain as rapidly and efficiently on rations with urea as with natural protein supplements. There is no significant evidence that any differences result in carcass cutability or dressing per cent.

Select Most Economical Protein Feeds—There are many satisfactory protein-rich feeds that will properly balance a ration for beef cattle; therefore, it is desirable to select feeds for their protein content. The prices

MISSOURI 40% SUPPLEMENT*

Ingredient	Per Cent
Urea (262% protein equiv.)	10
Alfalfa meal	20
Ground shelled corn	60
Soybean oil meal	10
Total	100%
Vitamin A per pound	20,000 IU

*Plus feed additives in sufficient quantities to meet recommended daily amounts.

paid for protein-rich feeds should be based on the protein contained in them.

3. Providing Satisfactory Minerals and Vitamins

Whether or not minerals and vitamins must be supplied depends upon the feed and water the cattle are consuming. This in turn depends upon the geographic area in which the cattle are grazing or where the feed was produced, because the crops grown on soils are a direct reflection of the mineral content of the area and the water utilized by the plants. Rain water is low in minerals whereas well water from soils rich in lime would be high in calcium. Do not purchase minerals if the ration is already well supplied. The one exception is that of salt which must always be added. However, supplemented minerals and vitamins are needed when not supplied by pastures and other feeds. In many areas it is necessary to keep a mineral mix available at all times.

Salt should always be provided for cattle as they will eat from 1 to 2½ pounds per head per month and more on green grass. It is a desirable practice to distribute salt to encourage salt consumption and uniform grazing.

On dry grass block salt is a desirable form in which to feed salt. However, if cattle are grazing largely on succulent feed, it is a desirable practice to feed salt half ground as they can consume it more easily.

The minerals most apt to be lacking, in addition to salt, are calcium and phosphorus. A practical mineral mix is:

Equal parts: Salt—half ground
Phosphorus supplement (bone meal)
Calcium supplement (limestone)

If the area definitely is deficient in one or the other mineral, the percentage of one lacking should be increased.

Watch for Grass Tetany—Tetany is caused by a lack of magnesium which is accentuated by a calcium deficiency. Death may come quickly if a cow does not receive treatment. Cows with calves are most prone to be affected. A cow may appear well in the evening yet be dead the following morning.

The disease appears most often during the winter months when pasture grasses growing under adverse conditions seem to lack adequate amounts of minerals. Dry lot animals seem to be less susceptible. Therefore, it is a desirable practice to move cows and calves to dry lot feeding conditions if the disease strikes or is suspected. Many livestock men believe that a mixture of magnesium oxide, bonemeal and trace mineral salt in equal proportions will prevent an outbreak during the unfavorable growth period for pasture. A desirable practice would be to supply this mixture during winter months. Place the material out early in the year or several months prior to the danger period so the cattle will establish a reserve of magnesium.

If the disease strikes, a veterinarian or livestock specialist from the Agricultural Extension Service should be contacted immediately. Cows can be saved by an injection of mineral (calcium and magnesium) under the hide or into the vein. High magnesium salt blocks are available commercially.

Supply Minerals According to Specific Area

Deficiencies of small amounts of other minerals, such as iodine or cobalt, must be supplemented for in the diet in certain areas. In other areas there may be a definite toxic effect because of too much of a certain mineral, for example, fluorine. Local authorities should be consulted in order to detect and properly correct such deficiencies.

Guard Against Mineral Excess

Other minerals may be involved but molybdenum is a case in point. Certain soils contain an excess of the element. Plants, especially legumes, growing in irrigated pastures, contain amounts toxic to cattle. Young cattle are affected most as evidenced by severe scouring and loss in weight.

Molybdenosis should be suspected in unthrifty cattle which show these symptoms: persistent scouring, rough hair coat, dehydration, arching of the back, listlessness and weakness, stunted growth or emaciation. Sometimes animals show lameness. In many cases there is a change in hair color from black to mouse gray or from red to bleached-out straw color. Death is rare but retarded gain and growth are common.

University of California Extension Service recommends small amounts of copper sulfate daily as treatment for cattle on such feeds. It may be mixed in the feed or placed in the drinking water. The rate is one ounce of copper sulfate to 250 gallons of water. Do not overdose. In severe situations, or as a preventative, copper glycinate can be injected. Consult a veterinarian regarding dosage and administration of the compound.

Provide Necessary Vitamins—Sunshine provides vitamin D. Therefore, cattle should have access to open areas whenever weather permits. Grazing green grass usually supplies the vitamins cattle need. However, should they be forced to eat dry grass entirely, they will gradually become deficient in vitamin A. Under these conditions it is a desirable practice to feed 8 to 10 pounds of bright green hay every 3 or 4 days.

4. Providing Sufficient Water

Provide Ample Clean Water—It is important that cattle have access to clean water at all times. Mature cattle will need about 10 gallons per head daily. Water should be uncontaminated so as to prevent the spread of disease. If the weather is extremely cold and freezing, the water should be warmed slightly.

Properly Distribute Water—On large pastures and under range conditions, sources of water should be distributed in order to promote uniform grazing and complete utilization of feed in remote corners of the range or pasture. At the same time, watering places should not be so far apart that the comfort of animals is at risk or that too much weight is walked off in going and coming from watering troughs. If animals have to go too far for water, they will not take the trouble and consequently do not drink enough to meet their needs.

5. Preparing Feed Properly

There is no single answer as to how feeds should be prepared. This will depend on kind and quality of feed, cost of preparation, and kind of stock consuming the feed.

Grind Small Grains—Small hard grains, such as rye, barley, wheat, oats, and grain sorghums, should be ground or rolled for cattle feeding. Coarse grinding is generally preferable to fine.

Grind Corn and Cob Meal—Iowa State Extension Bulletin, page 99, states that cost of grinding corn and cob meal is little different from that of shelling. Experiments show that corn and cob meal is safer and goes somewhat farther than shelled corn. Corn and cob meal is an excellent way to start cattle on feed, as there are few digestive disorders. This feed is best for home conditions, as rapid gains are not so important as with the commercial feeder.

Do Not Grind Shelled Corn—If hogs are getting the "pick up," there is no saving in grinding any but very hard shelled corn and fine grinding is worse than none. It slows corn consumption, hinders digestion, and wastes feed. In addition, hogs do not get the "pick-up."

Other Methods of Preparing Concentrates—As a general rule, cooking, soaking, sprouting, or fermenting does not increase the food value of feeds for beef cattle and is an uneconomical practice. Occasionally those fitting animals for show utilize one or more of the above practices as a means of whetting the animals' appetite and increasing total consumption.

Do Not Grind Roughages—It rarely pays to grind hay but if it is ground, it should never be ground fine.

Shred or Chop Poor Quality Roughage—It is highly questionable if it pays to chop high-quality hay. When a high quality hay is properly fed in suitable racks, there is little waste. One exception to not chopping

Fig. 5.7—Concrete trough which provides ample clean water. Note how automatic float is protected from livestock.

high quality roughage is in the use of new type mechanical field choppers that are used as part of the normal hay harvesting operations. In this manner, there would be no extra cost for chopping; yet the producer has the advantage of increased storage capacity and less feed wastage of chopped hay when fed.

In the Southwest even mesquite has been successfully fed to cattle by chopping. Other areas report various kinds of native brush ground and fed to cattle.

Prepare Silage—According to Iowa State University Bulletin, more pounds of beef can be obtained from an acre of corn fed as silage than an acre fed in any other way. Many other feeds and even grass can often be profitably made into silage. The use of trench silos may cut down the amount of capital invested in storage equipment. Mechanical equipment has taken much of the high labor drawbacks out of making and feeding silage.

Add Molasses to Poor Quality Roughages—In areas where molasses can be reasonably obtained, palatability and consumption of poor quality roughage can be increased by addition of molasses. The material must be thoroughly mixed by means of mechanical mixers at the time of chopping. In addition, the dust problem is cut down somewhat by the addition of molasses.

Feed Cubes—One of the most revolutionary breakthroughs in livestock feed preparation has been the development of the hay cube. Hay cubes are dense, bite-sized morsels of hay which are field-produced by machines picking up hay from the windrow, or by stationary machines. Cubes have completely changed handling and feeding methods for alfalfa hay. Cubed hay has been tested and accepted by livestock feeders because of its improved density, better flow and bulk-handling characteristics, reduced storage space, increased feed consumption by animals, and labor savings.

Cube Size and Shape—Many sizes and shapes of cubes have been produced by various experimental machines, but cubes now available commercially are 1¼ inches square and vary from 1 to 4 inches in length. This size allows optimum feeding, has acceptable ma-

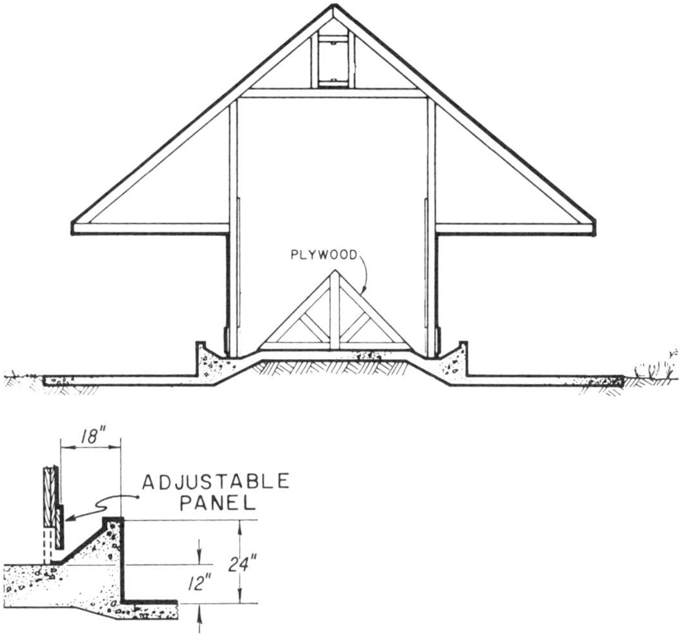

Courtesy, Circular 550, California Agricultural Experiment Station

Fig. 5.8—Above: Alternate barn-floor design to eliminate handling cubes in lower part of storage. Note reduction in storage space. Below left: Recommended manger cross-section for a self-feeding cube barn. Sloping bottom reduces build-up of fines in the manger. Adjustable panel may be lowered to stop or control flow of cubes and force animals to eat fines.

terials-handling characteristics, and can be produced at reasonable capacity with a farm-size power unit.

Moisture Content—Alfalfa hay should be cured to about 10 per cent moisture content before cubing. Water is sprayed on the windrow of hay as it enters the cubing machine, resulting in warm cubes at 14 to 16 per cent moisture as they emerge from the machine. For safe storage, cubes should be cooled and dried by natural or forced aeration to 14 per cent or less mois-

Fig. 5.9—Loading cubes. One of the many advantages of hay cubing is that it completely mechanizes the haying operation. This loader can scoop up over one ton of cubes in a single operation.

ture content. Excess moisture and heat in stored cubes cause mold growth and loss of quality and nutrients; they also can result in loss of crop by fire.

Little Waste—There is practically no waste when feeding cubes as there is with loose or baled hay. For this reason cattlemen can afford to pay considerably more for cubes than hay prepared by any other method.

Accustom Cattle to Cubes—Generally cattle eat cubes readily right from the start, although at times they will shun it in favor of feed they are in the habit of eating. Spread a few cubes alongside or on top of their usual feed and in a few days they will readily consume cubes. Provide plenty of fresh water, as large amounts of it are consumed when cattle are eating cubed hay.

6. Feeding the Beef Calf

There is no substitute for a good mother and the ample supply of milk she furnishes in properly raising a beef calf. Not only does mother's milk furnish a complete, well-balanced diet, but it is available to the calf at all times at the proper temperature thereby eliminating to a large extent digestive disorders. No labor for the rancher is involved in feeding, and the calf gets his food under extremely sanitary conditions. However, if the dam is a poor milker or feed conditions are inadequate so that her own body suffers as well as her offspring, then the food the calf receives should be augmented from sources other than his dam.

Fig. 5.10—Creep feeding of calves on pasture.

Creep Feed Calves—If natural conditions are unfavorable or their dams are heifers or inferior milkers, calves should be creep fed. When calves are creep fed from a few weeks of age, the total feed required to fatten them is generally less than if they are not fed until after weaning. In addition, creep feeding during the nursing period produces more uniform calves.

The following ration for creep feeding calves is recommended:

3 parts by weight	Cracked corn—equal parts Crushed oats—equal parts Wheat bran—equal parts
1 part by weight	Linseed meal

Keep Feed Available—Keep the mixture available to the calves at all times, as any sudden change in amount eaten may cause digestive troubles.

Continue to Feed Breeding Stock — Heifers and young bulls intended for breeding stock should be kept growing well throughout the entire year so as to attain full size. The expense of bringing calves to maturity may be increased if they are stunted while young.

Provide Hay or Pasture—Calves will begin to graze at an early age even though they do not consume large amounts. Green grass or good quality hay should be available to them.

Raising the Orphan Calf—Occasionally a cow will die during parturition. Quick action can save the calf. The orphan calf may be placed with a cow who has lost a calf or it may be nursed by a cow with another calf about the same age, providing the cow has good milk production. Some cows will refuse to take a strange

Fig. 5.11—Calf nursing its mother. There is no substitute equal to an ample amount of milk from the dam.

calf; however, if they are put together in a stall for a short period, the two will often become compatible. Under these conditions, supplemental feeding is a desirable practice.

Skin the Dead Calf—In order to make the foster mother take a strange calf, it is a desirable practice to skin the dead calf and tie the hide onto the back of the strange calf. Therefore, the cow that has just lost her calf will accept the different one as her own. The beef herdsman for the University of California beef herd states that this practice works well if the two animals are left together for several days in a small corral.

Keep a Nurse Cow—In purebred herds where orphan calves must be saved, it is a desirable practice to keep a dairy-type nurse cow on hand for such emergencies.

One good dairy-type nurse cow can easily raise two or more beef calves.

Raising Calves with Limited Milk

Feed Whole Milk for 10 Days. A general rule is to feed one pound of milk for each 10 pounds body weight; one quart weighs about 2 pounds.

Increase Amount of Milk Gradually. Feeding twice daily is sufficient but never to the limit of the calf's appetite.

Limit Amount of Milk. After 10 days the milk may be limited and changed to skim milk. Provide concentrates and alfalfa hay at this time. A good concentrate mix to be fed with limited milk is:

6 parts barley or oats
3 parts wheat bran or mill run
1 part oil meal
1 per cent salt

When milk feeding is discontinued, add 18 per cent dried skim milk or 9 per cent fish meal to concentrate mixture. Feed this mixture until 6 months of age. Always provide hay. At six months of age, barley and alfalfa hay are sufficient unless fattening for show.

7. Feeding the Breeding Herd

Feed Economically—Economical, efficient feeding of the breeding herd is possible through maximum use of pastures and roughages, supplemented with the necessary minerals, vitamins, and minimum amount of concentrates, which are generally protein.

Sort the Breeding Herd—It is a desirable practice to sort the animals into various groups according to

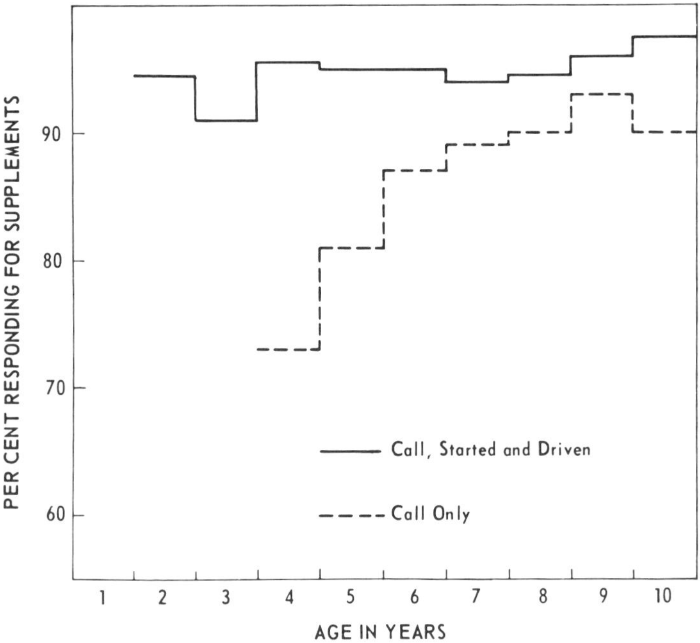

Courtesy, Bulletin 819, California Agricultural Experiment Station

Fig. 5.12—Effects of social dominance on response to call in mixed-age breeding cows. When efforts were made to get as many cows as possible to the feeding area by calling them, starting them on their way, and driving them (solid line), response was higher in all age groups, as compared with a system whereby cows were called only (dotted line). The difference between age groups was significant in each case. Two-year-olds were present only the first four years of an 11-year study period. During the five-year period when cows were called only, three-year-olds were not added to the herd until after they had calved.

sex, size, strength, and condition. This is especially true for winter feeding on the range. However, it must be consistent with good management practices and available fencing.

Get Rid of "Bossy" Cows—In the long run cattle will do better if the over-aggressive cattle are culled and a more manageable docile herd is developed. Feeding problems are exaggerated when some cows exert their dominance on the rest of the herd. Kenneth A. Wagnon, Specialist in the Agricultural Experiment Station, University of California, Davis, has studied the phenomenon at length and published an excellent report in Bulletin 819, "Social Dominance in Range Cows and Its Effect on Supplemental Feeding." This publication discusses the many factors such as age, weight, aggressiveness, timidity, and so on that influence behavior of cattle.

SUMMARY OF THE EFFECT OF PROTEIN SUPPLEMENTATION TO PREGNANT COWS ON A DRY GRASS RANGE (7 YRS.)

	Supplement	Control
Pregnancies, %	89.6	77.2
Calves weaned, %	83.0	67.6
Weaning wt., lb.	470	417
Calves per cow, lb.	390	282
Yearly supplement, lb.	365	18

Properly Feed Breeding Cows—If good green pasture is available or good legume hay, such as alfalfa or clover, constitutes half or more of the ration, there is little need for adding concentrates to the ration for mature cows. Younger or lighter weight cows that are on silage or cheap roughage should be fed from 1 to 2 pounds daily of cottonseed meal or a comparable protein concentrate. According to Farmer's Bulletin 1179, few combinations are more economical than a ration of corn silage, straw, and cottonseed meal for wintering beef cattle.

Feed Nursing Cows Heavier—Cows that are nursing calves should receive more protein supplement than dry cows. However, feed only the amount that is necessary to properly balance the other feeds which consist largely of roughages. Ordinarily beef cows raising calves are not fed more than 2 pounds of supplements each daily, and then only in the winter. Suggested rations for wintering 1,000-pound beef breeding cows:

Ration No. 1		Ration No. 2	
Corn silage	30 lbs.	Grass hay or stover	18 lbs.
Cereal straw or stover	Unlimited	Cottonseed cake or equivalent	2 lbs.
Cottonseed meal or equivalent	1½ lbs.		

Properly Feed the Herd Bull—Good roughage should make up most of the ration although some grain usually is needed to keep bulls thrifty and in good condition. Cornell Ext. Bull. 581 says bulls should be in good flesh, but not fat, when turned in with the breeding herd. Bulls on good pasture need no additional feed, although it is often a desirable practice to feed a protein concentrate, such as cottonseed, for about 60 days before the breeding season.

Regulate Supplements with Salt—In large herds on the range it is a problem to prevent "bossy" cattle or the stronger ones from eating larger amounts of concentrates than weaker animals who especially need supplemental feeding. This is a serious problem during severe winters. Many stockmen find it a desirable practice to mix cottonseed meal and salt on a 50-50 basis. Large amounts of salt apparently are harmless to cattle and an individual animal will consume a maximum of around 1 pound of salt daily, thereby consuming 1 pound of protein feed, and then stop eating for that day so that weaker cattle will have an opportunity to eat. The percentage of mixture can be varied so as to automatically feed the desired amount of supplement per head per day; for example, 100 pounds of meal and 50 pounds of salt would provide 2 pounds of meal per cow per day. Plenty of drinking water must be available.

Space Feedings Week Apart—Common practice in winter supplemental feeding is to feed a small amount each day. Many livestock men believe better results are obtained if several days or even up to one week elapses between feedings. Reasons for this practice are that cattle will "rustle" better for feed between feedings, as they tend to wait for the feed truck if fed every day rather than search for native feed. Another advantage is that dominant cows will drive away weaker ones if only a small portion is fed daily whereas if larger amounts are fed, say every three or four days, every animal tends to get the proper amount.

Feed Two Kinds of Hay—Most livestock men believe that cattle do better if more than one kind of roughage is available. In particular it seems to cut down on simple scouring. This is especially true if cattle are on green pasture and some type of dry feed can be provided. Whenever animals are feeding continually on a feed like alfalfa hay they seem to relish a change to another kind of hay even if it is an inferior feed such as straw, stubble, or native grass hay.

8. Feeding Cattle for Fattening

Fattening Advisable—While fattening cattle is not an economical procedure for every ranch, it is a desirable practice for the industry as a whole. Fattening increases tenderness and palatability, protects the carcass from spoiling while aging, and decreases shrinkage. Moreover, it raises the dressing percentage and improves appearance of the product. From the farmer's standpoint, this means added profit. However, even if it were not for this factor, farmers who raise large amounts of concentrates would still fatten cattle in order to maintain fertility and provide an outlet for low-cost roughages, such as hay or silage, which at times would have no market except through livestock.

Fit the Cattle to the Feed Supply—Farms differ a lot in the amount and kind of feed they have for a cattle feeding operation. Some farms have mostly corn. Other farms have lots of hay, but little corn. Still others have pasture that has little use except for a cattle feeding program. In some years there may be an extra supply of lower grade roughage to market. In order to make the most profit, it is essential that the amount and kind of cattle be consistent with the feed program.

Watch the Outlook—It is a "must" for the cattle feeder to keep up to date on the business outlook. This is twice as important for the buyer of heavier cattle. A good rule is when the outlook is bearish, to buy cattle late, buy them light, and don't buy too many. The reverse is true when the market is bullish. Of course, it is easier to say than do. Most cattle feeders do better if they buy their cattle at the time of year when stockers and feeders are usually cheapest. Since range men must sell most of their cattle at the end of the pasture season, stockers and feeders are usually cheapest at that time.

Purchases of stocker and feeder cattle bought by Corn Belt farmers from out of state are distributed by seasons as follows:

	From each 1,000 head
September to November	530 head
December to February	168 head
March to May	152 head
June to August	150 head

Prices are generally lower when the greatest number are offered for sale.

Select the Right Feeders—The higher the grade or quality of feeder, the more economical will be its use of feed. Farmer's Bulletin 1549 states the high-grade steer has greater capacity for feed and makes more economical use of it in laying on fat in the region of the valuable cuts. It is a desirable practice for farmers to let reputable buyers purchase their feeders for them unless they are highly skilled themselves and have reasonable access to supplies of cattle. The quality and nature of available feeds and the length of feeding period should largely determine the age of cattle to feed. The younger the animals, the longer will be the feeding period. Older cattle fatten in a much shorter time.

Cattle feeders usually prefer thin cattle for the feed lot. When cattle are relatively cheap and feeds high, cattle possessing considerable fat are preferable. Feeders in good flesh are sought for short feeding periods.

Heifers vs. Steers—Steers make slightly more total gains than heifers. Heifers, however, fatten more rapidly than steers. Spayed heifers do not gain more economically than open heifers nor do they excel in dressing percentage. In quality and palatability, heifer beef is equal to steer beef and in many parts of the world is preferred.

Select Desirable Ration—It is a desirable practice to select an adequate fattening ration, always keeping in mind that utilization of the cheapest feeds without reducing the effectiveness of the ration is the goal that will yield the greatest profit.

Following are suggested rations:

TYPICAL RATIONS MEETING REQUIREMENTS

Growing 300-400 lb.	8 lb. Alfalfa hay 2 lb. Barley	3 lb. Alfalfa hay 6 lb. Sudan hay 1 lb. Corn	4 lb. Oat hay 2 lb. Cottonseed meal 3 lb. Almond hulls
	10 lb.	10 lb.	9 lb.*
Growing 500-600 lb.	10 lb. Alfalfa hay 2 lb. Wheat 4 lb. Barley straw	8 lb. Alfalfa hay 20 lb. Corn silage (7 lb. Dry matter)	7 lb. Sudan hay 4 lb. Almond hulls 2 lb. Cottonseed meal 2 lb. Molasses
	16 lb.	28 lb. (15 lb. D.M.)	15 lb.*
Finishing 700-800 lb.	8 lb. Alfalfa hay 12 lb. Barley	6 lb. Oat hay 10 lb. Corn 2 lb. Cottonseed meal 2 lb. Beet pulp	3 lb. Alfalfa hay 3 lb. Barley straw 7 lb. Wheat 5 lb. Barley 1 lb. Molasses-urea mix
	20 lb.	20 lb.*	19 lb.*
Finishing 900-1000 lb.	6 lb. Alfalfa hay 6 lb. Oats 8 lb. Corn 2 lb. Beet pulp	7 lb. Alfalfa hay 2 lb. Oat hay 7 lb. Barley 8 lb. Corn 1 lb. Molasses	2 lb. Alfalfa hay 2 lb. Sudan hay 2 lb. Beet pulp 16 lb. Barley
	22 lb.	25 lb.	22 lb.*

*Use trace mineralized salt, free choice.

Fig. 5.13—Cutting the tips of a steer. Dehorned animals do better in the feedlot. Occasionally horns of some feeder cattle may be tipped in order to subdue them for a brief feeding period or to control an individual animal.

Where large amounts of corn are available, Farmer's Bulletin 1549 suggests the following rations:

FEEDING CATTLE FOR BEEF

400-POUND STEERS

Dry Rations	Pounds	*Succulent Rations*	Pounds
Corn	13	Corn	10
Legume hay	6	Protein meal	1¾
		Mixed hay	3
Corn	10	Silage	7
Protein meal	2		
Mixed hay	5	Corn	10
		Protein meal	2
Milo, kafir, or barley (ground)	12	Straw or stover	3
Protein meal	1¾	Silage	10
Sorgo fodder	8		
		Milo, kafir, or barley (ground)	10
		Protein meal	1¾
		Grass hay	3
		Silage	15

600-POUND STEERS

Dry Rations

	Pounds
Corn	16
Legume hay	7
Corn	14
Protein meal	1½
Mixed hay	6
Milo or kafir (ground)	12
Protein meal	2½
Cottonseed hulls	10

Succulent Rations

	Pounds
Corn	12
Protein meal	2
Mixed hay	4
Silage	9
Corn	10
Protein meal	1½
Legume hay	5
Silage	20
Milo or kafir (ground)	12
Protein meal	2¼
Sudan hay	4
Silage	15

800-POUND STEERS

Corn	18
Legume hay	9
Corn	15
Protein meal	2
Mixed hay	8
Milo or kafir (ground)	15
Protein meal	3
Cottonseed hulls	15

Corn	14
Protein meal	2¼
Mixed hay	4
Silage	14
Corn	12
Protein meal	2¾
Straw or stover	3
Silage	20
Milo or kafir heads (ground)	15
Protein meal	2
Grass hay	3
Silage	15

1,000-POUND STEERS

Corn	20
Legume hay	10
Corn	16
Protein meal	2½
Mixed hay	10
Corn	18
Protein meal	2½
Straw	5
Stover	5

Corn	18
Legume hay	5
Silage	25
Corn	16
Protein meal	2½
Mixed hay	4½
Silage	17
Corn	15
Protein meal	2½
Straw or stover	5
Silage	30

Courtesy, U.S.D.A.

Follow Cattle with Hogs—When fattening cattle are being fed corn, corn silage, and other feeds with whole grains, hogs should follow them to consume the undigested and spilled grain. At least one 100-pound hog should be provided for each 1,000-pound steer.

Start Animals on Feed Correctly—If an animal does not get started off correctly, it is almost certain to be an unprofitable animal. Starting the feeders on very limited quantities of concentrates and increasing these gradually during the first thirty days is essential.

Farmer's Bulletin 1549 suggests the following method of increasing feed in the rations for fattening various weight steers:

One "Best" Ration—There are many rations which will fatten cattle; however, the producer should determine the one that most nearly meets his needs. Consideration must be given to home-grown feeds and other economy factors as well as the nutrition of his cattle. Rations should not be changed unless there is a reason and then very gradually over a period of one or two weeks.

SUGGESTED METHOD OF INCREASING FEEDS IN THE RATION FOR FATTENING STEERS OF VARIOUS WEIGHTS

Ration and weight of steers	Daily feed per head for first day and at beginning of stated periods										
	First day	Second week	Third week	Fourth week	Second month	Third month	Fourth month	Fifth month	Sixth month	Seventh month	Eighth month
Corn and legume-hay ration:[1]											
400-pound steers—	*Lbs.*	*Lbs.*	*Lbs.*	*Lbs.*	*Lbs.*	*Lbs.*	*Lbs.*	*Lbs.*	*Lbs.*	*Lbs.*	*Lbs.*
Corn	3	5	7	8	10	12	15	16	16	15	......
Hay	5	6	6	6	6	6	6	6	6	5	......
600-pound steers—											
Corn	4	6	8	10	14	16	20	20	18	......	......
Hay	6	8	8	7	7	7	7	7	6	......	......
800-pound steers—											
Corn	5	7	10	13	18	20	22	20	......	......	......
Hay	8	10	10	10	10	9	8	8	......	......	......
1,000-pound steers—											
Corn	6	8	12	16	20	24	24	......	......	......	......
Hay	10	12	12	10	10	10	8	......	......	......	......
Silage ration:											
400-pound steers—											
Corn	3	4	4	5	7	9	10	11	12	13	14
Protein meal	¼	½	¾	1	1¼	1¼	1½	2	2	2½	2½
Hay[2]	3	3	3	3	3	3	3	3	3	3	3
Silage	6	8	8	8	8	8	7	7	7	6	5
600-pound steers—											
Corn	3	4	6	8	10	12	14	14	14	14	......
Protein meal	½	¾	1	1¼	1¼	1½	2	2½	3	2¾	......
Hay	4	4	4	4	4	4	4	4	4	4	......
Silage	8	10	10	10	10	10	10	9	8	6	......
800-pound steers—											
Corn	4	6	8	10	12	16	16	16	16	......	......
Protein meal	½	1	1¼	1½	1¾	2	2½	3	3	......	......
Hay	4	4	4	4	4	4	4	4	4	......	......
Silage	10	15	16	16	16	16	14	12	10	......	......
1,000-pound steers—											
Corn	5	7	8	12	18	18	18	18	......	......	......
Protein meal	½	1¼	2	2¼	2½	2¾	2¾	2½	......	......	......
Hay	4	5	5	5	5	4	4	4	......	......	......
Silage	10	20	20	20	20	18	15	12	......	......	......

[1]Legume hay, such as alfalfa, clover, soy bean, or cowpea.
[2]The hay in silage ration may be a mixed hay or any of the grass hays.

Courtesy, U.S.D.A.

Be Watchful for Feeding Difficulties—Good cattle feeders are continually alert for the first sign of an animal going off feed. The ration should be critically examined for quality and amount being fed so as to eliminate the difficulty. Generally, reducing the concentrate part of the total amount fed will assist cattle to resume normal digestive processes.

Typical symptoms of cattle going off feed are that they will nose feed around in trough and not eat at feeding time; feces become more "loose" and begin to approach diarrhea condition; animals become listless and stand around with a dull expression in their eyes. Symptoms are progressive and the feeder should be watchful to catch the first sign.

Cattle going off feed is a serious proposition; therefore rations should be changed only when necessary and there is a reason. If changes must be made, it should be done gradually and over a period of several weeks.

Utilize Grass Pasture—Whenever possible, pasture should be the foundation for feeding steers since gains made on pasture are usually the cheapest. Corn or cereal grain, such as barley, should supplement the last half of the feeding ration. It is important that pastures be in good shape and not weedy or full of woody plants for best fattening results. Legume pasture, such as Ladino clover or alfalfa, is excellent for fattening cattle. Ample salt must be provided on such succulent feeds. Pastures should be growing well and partially mature before turning in cattle so that they can get all they want to eat without working too hard. This practice is also beneficial to the pasture.

Provide Some Coarser Roughage—In order to reduce the hazard from bloat, some coarse feeds, such as dry hay or even straw, will greatly assist in preventing bloat. This should be available to cattle at all times while grazing green legume pasture. Cattle should never be turned onto lush pasture when their stomachs are empty.

Effects of Stilbestrol with Steers—Stilbestrol, the synthetically produced hormone, is an estrogen very similar to the female sex hormone normally secreted by the ovaries. Treated cattle gained ½ to 1 pound more a day than untreated.

Stilbestrol Is Banned—In 1972 the Food and Drug Administration issued Federal regulations which require cattle and sheep producers to withdraw feed containing diethylstilbestrol *seven* days prior to slaughter, and certify in writing the seven-day withdrawal period has been observed.

No residues of DES are permitted in meat.

On July 31, 1972, the Federal Drug Commissioner said DES could no longer be added to feed. There will be many reactions and results to the discontinuance of this feed additive, but most certainly the cost of producing meat will increase. *Implants* (ear implants) *were permitted* although rulings by the Federal Drug Commissioner in 1973 banned even this form of stilbestrol.

Keep Abreast of Regulations

New compounds will undoubtedly be developed in an effort to increase the rate of gain and lower the cost of production of beef. These could be very beneficial to the livestock industry, so producers should keep up to date on what is available and what can be safely and legally used.

Zeranol, for example, is a new chemical compound used for implants marketed under the trademark of **Ralgro.** It is an anabolic agent for growth promotion.

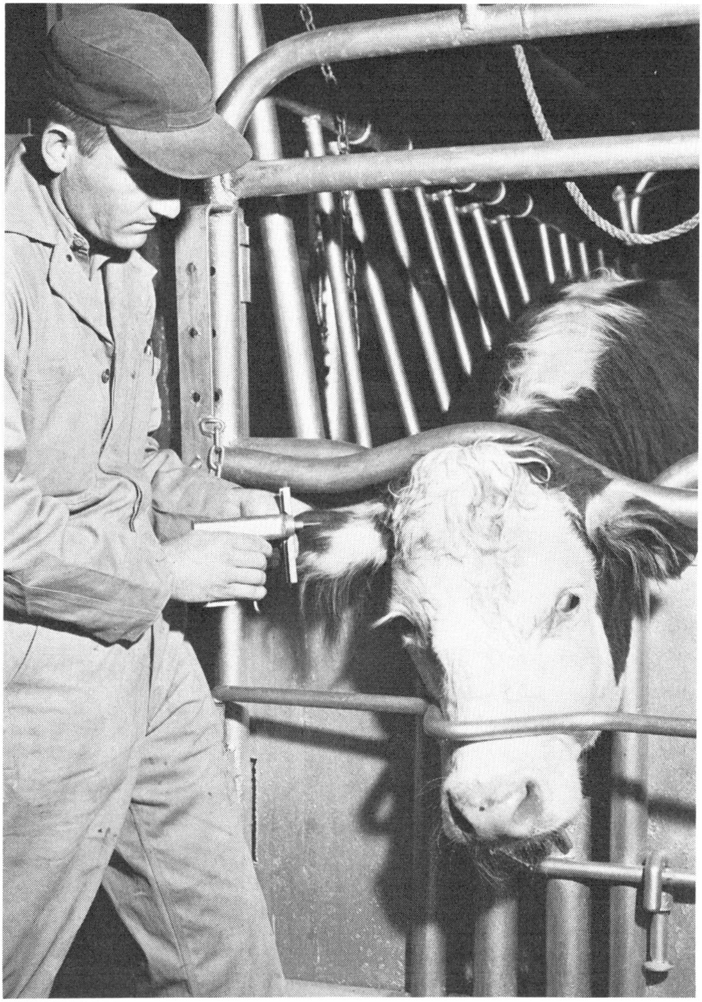

Fig. 5-14—Implanting a stilbestrol pellet in the ear of a beef steer. In addition to showing the use of an implant gun, other desirable approved practices are shown in this picture including a desirable type steer properly dehorned as a calf and a portable squeeze chute to handle livestock safely and quickly. (All forms of stilbestrol including implants have now been banned by the FDC.) Other kinds of implants such as Ralgro, are available.

Suggestions and Rules of Thumb for Feeding

a. Introduce the cattle to concentrates gradually. It should take at least three weeks to get the animal on full grain feed. Segregate animals as to size.

b. On full feed, an animal will consume from 1.5 to 2.0 lbs. of concentrates per 100 pounds live weight. This figure is somewhat high unless animals are being fitted for show.

c. Start calves on 20% concentrates and 80% roughages. This can be increased to approximately 60% concentrates and 40% roughages.

d. Pasture is an excellent source of feed, and animals can gain from $\frac{3}{4}$ to $1\frac{1}{4}$ lbs. a day on pasture. On full grain feed, they gain from $1\frac{1}{2}$ to $2\frac{1}{2}$, depending on size and feed.

e. Feed should be ground medium-coarse or rolled. Fifteen to 20% beet pulp is excellent to add bulk to the ration.

f. Wheat and milo may make up 30% of the concentrate mix. Oats and wheat bran may make up 15% of concentrate mix.

g. Cottonseed, linseed, and soybean meal are about equal in value and should be used to supplement low protein rations. However, it is not needed when cattle weigh 650 to 700 pounds and are in thriving condition with alfalfa as the only roughage.

h. Cattle eat about 1 pound hay per 100 pounds live weight.

i. Three pounds silage equal 1 pound hay, so feed 3 pounds silage for each 100 pounds live weight.

j. Don't change the ration abruptly.

k. Don't turn cattle on luxuriant clover when the dew is on. Provide some coarse roughage to prevent bloat.

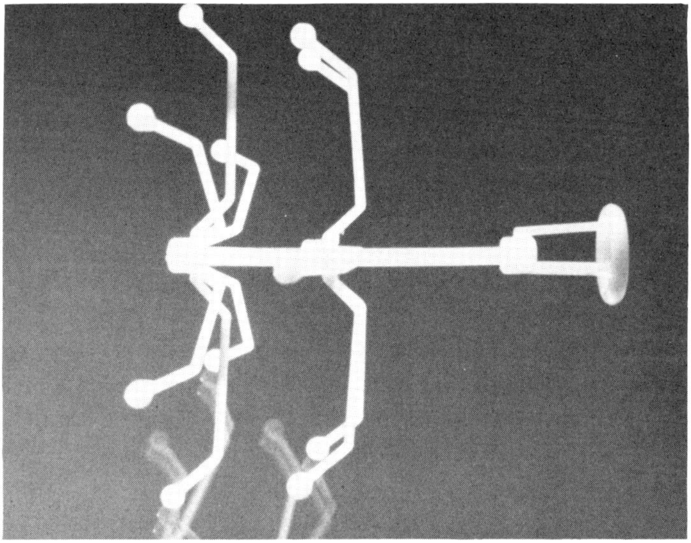

Fig. 5.15—The Hei-Gro Device. This small gadget is a non-hormone, non-drug gain stimulant which reduces riding and keeps beef heifers quiet. It is a molded, food-grade nylon device that is inserted into the forward portion of the heifer's vagina touching the cervix. The pressure of the device tends to stimulate the cervix and vagina walls. It leaves no residue. It needs no replacement, removal, or withdrawal for the life of the heifer.

1. Utilize good feeding equipment to save feed and labor.

Use Antibiotics with Poor Doers—A great deal has been said about the merits of antibiotics like terramycin or aureomycin. Most authorities seem to agree that little is gained by using these compounds if animals are raised under healthy, parasite free conditions and receive a proper diet. On the other hand if stock are not doing well and seem to be "ailing" or are not in a favorable environment excellent results are often obtained by including minute amounts of these substances in the ration. Mixing is a problem so the practical solution is to obtain them in commercial feeds and use according to manufacturer's directions.

Young stock or cattle with parasite problems seem to respond better than mature cattle. Perhaps the most desirable program is to feed and manage cattle properly so that antibiotics are unnecessary, using them only under special conditions as described.

9. Feeding Beef Calves for Show

Select Outstanding Calves—Select fancy or choice calves from a herd which is known to produce good feeders. In order to produce a really top-notch calf, it is often necessary to start with an animal that might qualify as a herd sire had it not been castrated. From an economic standpoint, this practice can seldom be justified, as such animals are too high priced. Young people who wish to exhibit in the Junior Division of livestock contests can obtain valuable experience and pleasure by selecting feeder calves that are not exceptionally high priced. Even calves in this classification make outstanding individuals at times, and thereby test and challenge the ability of a boy or girl to properly fit a beef animal for showing.

Fit More Than One Calf—Single animals often are restless and seldom "do" as well as several calves together. Equipment and labor are also more economically utilized if more than one calf is fed out.

Start Calves on Feed Slowly—Calves cannot adjust themselves quickly to a change of feed. It may take two to six weeks to accustom them to the grain ration. Agricultural and Mechanical College of Texas Cir. 226 recommends the following feeding schedule for weaned calves 400-500 pounds weight:

Feeding Schedule for First 15 Days
—Preliminary Feeding Period

First 3 days—All the hay calf will eat.

Second 3 days—½ pound cottonseed cake per head per day. All hay calf will eat.

Third 3 days—1 pound cottonseed cake per head per day. All hay calf will eat.

Fourth 3 days—1½ pounds cottonseed cake per head per day. All hay calf will eat.

Fifth 3 days—2 pounds cottonseed cake per head per day. All hay calf will eat.

Feeding Schedule Second 15 Days
—Preliminary Feeding Period
Grain Mix

30 lbs. cracked yellow corn or ground milo grain
25 lbs. whole or crimped oats
5 lbs. alfalfa leaf meal

Protein Supplement	Roughage
30 lbs. cottonseed cake	100-150 lbs. cane and Johnson grass hay

Keep the cottonseed cake at 2 pounds per day and slowly add the grain mixture by hand feeding the calf.

Hand Feed the Calf—Hand feed the grain mix at the rate of one-half pound per head per day and make a one-half pound increase each day per head for next fifteen days. By hand feeding, the calf can be closely observed for any digestive disorder and the feed reduced accordingly.

Self-Feed the Calf—After the preliminary feeding period and the calf is on full feed, the following mixture can be supplied in a self-feeder:

(100 lbs. mixture)

1st 30 days	**4th 30 days**
50 lbs. cracked corn	65 lbs. cracked corn
25 lbs. whole oats	15 lbs. whole oats
20 lbs. cottonseed screenings	15 lbs. cottonseed screenings
5 lbs. alfalfa leaf meal	5 lbs. alfalfa leaf meal

2nd 30 days
55 lbs. cracked corn
20 lbs. whole oats
20 lbs. cottonseed screenings
5 lbs. alfalfa leaf meal

3rd 30 days
60 lbs. cracked corn
20 lbs. whole oats
15 lbs. cottonseed screenings
5 lbs. alfalfa leaf meal

5th 30 days
70 lbs. cracked corn
15 lbs. whole oats
10 lbs. cottonseed screenings
5 lbs. alfalfa leaf meal

6th 30 days
75 lbs. cracked corn
10 lbs. whole oats
10 lbs. cottonseed screenings
5 lbs. alfalfa leaf meal

REMEMBER!
FEED YOUR CALF ALL THE HAY IT WILL EAT EVERY DAY.

From Texas A. & M. Circular 226

Use Most Available Feeds—Many locally grown feeds other than the ones suggested in the above rations will be just as acceptable in properly fitting a **show calf** and will reduce the cost materially.

Use a Nurse Cow—Unweaned calves must be put on a nurse cow if their dams are not available. Properly selected good nurse cows will let any calf suck so that the transfer to her of a strange calf should be no problem. Occasionally the nurse cow must be stanchioned while the new calf nurses along with her own calf until she becomes accustomed to the change.

Weigh Calves Regularly—In order to keep a check on the progress the animals are making, calves should be weighed every three or four weeks. This is a desirable practice even for commercial feeders in order to determine the progress cattle are making. However, under commercial conditions it is not necessary to weigh all animals, as representative samples will give a reasonable indication.

Brush and Gentle Calves—Calves respond to gentle, kind treatment. Brushing makes the hair lustrous and keeps the skin healthy. Calves that receive a reasonable amount of attention are not so apt to become nervous when strangers come around.

Keep the Calf Comfortable—Proper equipment, pens, protection, and management are of prime interest in keeping the calf comfortable. Fresh, clean water should be available at all times as well as granulated salt.

Suggested Rules of Thumb for Fattening Show Calves

a. Single animals do not do well alone. Better to have at least two and if possible, more.

b. Beef calves should be handled slowly and carefully for best results and gains. Don't practice your roping and riding on your fattening calves.

c. Animals should have access to dry, shady, and protected areas. These need not be warm, as animals create their own heat.

d. Animals that go to shows should be in a shaded, cool place in the day and turned out at night during warm weather. This helps to prevent shedding.

e. Exercise is important. Provide a large enough area for the animal to get sufficient exercise or else walk the animal daily.

f. Lice, ticks, flies, and warbles irritate cattle and prevent proper gains. Treat infected animals immediately.

g. To train animals, get them gentle first. Feed them in small, confining pens and gradually get the animals used to you and to being touched. Daily brushing and handling is essential for all animals that go to shows.

h. Animals may be trained to lead by tying them up and leading them to water two or three times a day.

i. The important thing is never to get excited or excite the steer. Careful, gentle handling will pay off in added pounds of beef at market time.

Fig. 5.16—Weighing a representative lot of calves to determine rate of gain.

10. Feeding and Testing Young Bulls

The growth of bulls is very important as the improvement in a herd probably relies more on the sire than any other factor. Extension animal scientists at the University of California, Davis, Roy V. Parker, James T. Elings, and W. James Clawson, have outlined the importance and practice of feeding growing bulls as follows:

The rate of gain is highly important in beef cattle. This gain in bulls should be natural growth, not just additional weight put on by heavy feeding, resulting in excess fat. Bulls should be fed enough to achieve their genetic potential for growth resulting in a strong, rugged condition at sale time. This fleshing condition would correspond to the Good and Low Choice grades in fed steers.

Also important is the heritability of rate of gain, useful in measuring performance of cattle for more precise selection. For postweaning gain, the heritability is approximately 40 per cent; for yearling weight-per-day-of-age (WDA), the heritability is 60 per cent, the highest there is for cattle.

Rations

Commercial feed mixes designed for this purpose are satisfactory although a home-mixed ration can be used.

If the grain ration is to be home-mixed, a simpler ration would be suggested:

	Per cent
Rolled barley or other grain	65
Beet pulp (dried molasses)	25
Cottonseed meal	10

In addition to this ration, provide free-choice trace mineral salt, bone meal, or di-calcium phosphate.

Roughage—Either hay or pasture should be of high quality. Bulls in a drylot should have alfalfa hay, a combination of alfalfa and cereal, or pasture hay.

Vitamin A—If hay or pasture is of excellent quality and good color, no additional vitamin A is necessary. If there is a question whether the roughage contains sufficient vitamin A, it can be included in the ration at 5,000 International Units (IU) per pound of concentrate. Or, bulls can be injected with vitamin A every 3 to 4 months at the rate of one million IU per head.

Daily Feeding Schedule

For growing bulls, concentrate rations should be fed at 1 or 1½ per cent of body weight. If the bulls are in good flesh, concentrates at 1 per cent of body weight should be adequate. If thin, they may be started at the 1½ per cent level and then reduced to 1 per cent.

Official Feed Test

Standardized feed tests can be conducted at central bull tests and also on the home ranch.

In feed testing for selection, it is recommended that bulls go on feed sometime following weaning but no later than a year of age. Breeders who wish to get bulls close to 1,000 pounds at a year of age should wean at 7 or 8 months and place calves on feed immediately following a short adjustment period after weaning.

Feed for 140 days if feeding at 1 per cent of body weight. Feed for 120 days of feeding at the 1½ per cent rate.

Length of Time on Feed

These rations can be used from weaning to sale time. They will put flesh on young bulls and allow the fulfillment of their genetic potential for selection.

After the test period, or when bulls reach 800 to 1,000 pounds, it should not be necessary to continue at the 1 or 1½ per cent feeding level. Level off at 8 to 10 pounds per head per day to keep them in good flesh for buyer appeal.

Management Considerations

Weighing—Individually weigh bulls at start and end of test. To obtain uniform weighing conditions among bulls, it may be necessary to take individual weights after an overnight stand without feed and water; or average two full weights, for each bull, taken on consecutive days at start and end of test. Group weights can be taken every 28 days to adjust the amount of concentrate to be fed.

Space—Provide at least two feet of trough space per bull and 200 square feet of pen space in a drylot.

Sorting—Sort either in sire groups or in mixed-sire groups by age.

11. Providing for Economical Use of Labor

One of the essentials for successful beef cattle production is the economical use of labor and equipment. This factor is especially important in the feeding part

Fig. 5.17—Conveniently arranged loading chute and cattle-handling corrals. Note use of barn as a holding pen.

Fig. 5.18—A well constructed feedlot. This feedlot is paved, has wire cable fencing to allow for ventilation, and has a unique shade arrangement. The rolls over the heads of the cattle are snowfences made of lath that can be rolled out over cable to provide shade at low cost in the summer and rolled up to let sun in during winter.

of beef production. The following suggestions should assist in accomplishing this:

Let Cattle Harvest Their Own Feed—Harvesting forage is perhaps the most labor consuming part of beef production. Full use of pastures where cattle harvest their own feed is a very desirable practice. "Cattling down" corn, or similar crops, is at times a desirable practice. This is especially recommended if the corn is soft or real late and hogs are allowed to follow.

Provide Cattle Guards—This prevents much useless time and effort in opening and closing gates. In addition, the gates then cannot accidentally be left open, as often happens, allowing stock to get out onto dangerous areas.

Provide Adequate Fencing—Not only should corrals and fences be well constructed, but they should be conveniently arranged so as to prevent useless motion.

Pave the Feed Lots—While cattle prefer to walk on

the ground and sometimes get foot sores on pavement, small areas that contain large numbers of cattle should be paved. Pave at least around the water tanks and feed bunks. Paved yards can be cleaned much faster and easier. If cattle must be held on such an area contin-

Fig. 5.19—Paved walk around the water trough. Troughs last longer and conditions are more sanitary if areas around water troughs are paved.

ually, it is well that they have access to an unpaved yard in dry weather.

Bulletin P99 from Iowa State University suggests the following in order to save labor in cattle feeding.

a. Convenient arrangement of feed storage, shelter, lots, and gates will minimize the time and labor required in feeding and handling cattle.

b. Have a constant and convenient automatic water supply easily accessible to all stock.

c. Haul, don't carry feed. Where numbers are small, the feed wagon may serve as temporary storage bin in the feed yard. Fill wagons directly from silo, bins, or hay storage.

d. Paved lots save labor, feed, and manure. Pave at least around water tanks and a 10-foot strip on each side of feed bunks.

Fig. 5.20—Natural shade provides excellent protection from summer heat.

e. Where cattle are being full fed, large capacity self-feeders permit handling feed in full loads with a minimum of labor. Self-feeding on pasture also saves hay making and manure hauling.

f. Feed hay close to storage or stack to provide the least possible handling.

g. Large-scale operators can make good use of elevating, conveying, and processing equipment.

h. Suitable pens, chutes, and stocks are great conveniences in sorting, vaccinating, spraying, etc., and are worth the cost to most operators.

12. Providing Comfort for Animals

Cattle can do without shade during summer heat, but always do better if shade is available. If natural shade cannot be obtained, simple open-side breezeways satisfy this requirement. In winter, artificial heating is unnecessary, but cattle should have access to areas protected against strong, cold drafts.

Spray for Flies—Spray with modern fly killers every two to four weeks depending on the season. Anything that increases the comfort of the animal will assist in making gains.

13. Protecting Cattle from Stock Poisoning Plants

Poisonous plants are the cause of heavy losses annually among all kinds of domestic animals on range or pasture. Injurious plants are found in all parts of the United States, but heaviest losses occur on western ranges. Larkspur, whorled milkweed, and locoweed are among the most common.

Purchase Good Quality Hay—Occasionally poisoning results from toxic plants being cured along with the roughage. Cattle are poisoned because they cannot segregate bad from good as they are able to do while grazing. Hay that is free from foreign matter and weeds would be less likely to contain toxic plants.

Keep Animals in Good Shape—Well-fed animals tend to be "choosy" while grazing and generally will ignore poisonous plants. University of California Bulletin 593 suggests the following precautionary measures:

Eradicate Poisonous Species—By use of chemicals, grubbing, or otherwise, poisonous plants should be eliminated from the pastures. If serious enough, well-defined areas should be fenced off.

Drive Animals Slowly—When herding animals from place to place, they should be moved slowly as they tend to devour anything within reach if rushed.

Keep Animals Well Fed—This has added significance if they are to be driven through an area containing poisonous plants.

Graze the Range with Proper Class of Livestock—Susceptibility to plant poisoning varies, so attention must be given to this factor. Cattle are seldom poisoned by death camas, whereas sheep losses are heavy. Sheep are probably never affected by larkspur, and cattle losses are heavy.

Provide Ample Salt and Minerals—Salt prevents animals from acquiring a perverted appetite which might cause them to consume poisonous vegetation. Free access to common salt is best.

Avoid Too Early Seasonal Use of the Range—Grazing too early in the season is a dangerous practice if the grass is so short the animals cannot obtain a good "bite," thereby eliminating the selection factor, as all plants tend to be the same size.

Sorghum Poisoning—This well-known domestic plant which includes Sudan and Johnson grass is well known for its lethal effect on livestock, including cattle, if conditions for poisoning are right. There is little danger if the following rules are observed:

a. Do not grow sorghums for pasture in an area where growing season is short and there is danger of late frost.

b. Pasture irrigated sorghums *after* they attain a height of 18-20 inches.

c. Do not graze sorghums immediately after a frost or drought.

d. Put only a few animals into field to test pasture if at all doubtful.

e. There is no danger from feeding dry or cured sorghums.

Expected Daily Gains for Steer Calves on Various Rations*

1. Full-feed well-preserved grass-legume silage plus 4 lb. grain — 1.2 to 1.5 pounds.
2. Full-feed well-preserved grass-legume silage plus 10 to 15 lb. grain — 2.0 to 2.4 pounds.
3. Full-feed corn silage plus 1.5 lb. supplement plus 4 to 6 lb. grain — 1.9 to 2.3 pounds.
4. Full-feed corn silage plus 1.5 lb. supplement plus 8 to 10 lb. grain — 2.2 to 2.6 pounds.
5. Full-feed good-quality mixed hay — 0.6 to 0.9 pounds.
6. Full-feed sorghum silage plus 1.5 lb. supplement plus 4 lb. grain — 1.3 to 1.7 pounds.
7. Average permanent pasture — 0.8 to 1.2 pounds.
8. Good improved pasture — 1.2 to 1.7 pounds.
9. Good pasture plus 1 lb. supplement plus 8 to 12 lb. grain — 2.0 to 2.5 pounds.
10. Full-feed grain plus 1.5 lb. supplement plus 3 lb. hay — 2.0 to 2.6 pounds.

---

*Leaflet 318, Extension Service, The Pennsylvania State University.

> Yearling and 2-year-old cattle usually make greater daily gains than calves; but the total amount of feed per animal does not vary much between different aged feeders which are of similar grade and finished to the same slaughter grade. To feed a steer to Prime, about 40 to 65 bushels of corn (or the equivalent in other feed grains) is required. The kind and amount of roughage determines how much grain is needed. To feed to the Choice grade, 20 to 45 bushels of corn should be expected. A Good slaughter grade requires 10 to 25 bushels. It is rather difficult to get cattle higher than a Standard slaughter grade with roughage alone.

Individual animals will often do much better than the figures above indicate.

Chapter VI

PROVIDING SHELTER AND EQUIPMENT FOR BEEF CATTLE*

Proper Equipment Essential—Proper shelter includes the necessary facilities depending on climatic conditions in order to adequately protect cattle from weather. Equipment includes watering and feeding devices, sorting and handling corrals, stocks, and other restraining equipment necessary to handle cattle rapidly and safely, both from the standpoint of safeguarding against injury to humans as well as to livestock. Most of this equipment must be constructed by the operator; however, desirable commercial equipment is also obtainable. Equipment should be designed and constructed so as to encourage the use of approved practices.

Activities Which Involve Approved Practices

1. Providing proper fencing.
2. Providing proper gates and cattle guards.
3. Providing proper handling corrals.

---

*A complete reference book, *Handbook of Livestock Equipment*, is available through The Interstate Printers & Publishers, Inc., Danville, Illinois. This book shows many plans for beef cattle not illustrated in this chapter.

Courtesy, Thuron Sprayer Mfg. Company

Fig. 6.0—No Bak cattle chute gate in action. The swinging gates at the rear of the lead cow permit cattle to move forward; but once past the gate, they are unable to back up. This speeds up movement of livestock. Equipment such as this lowers labor requirements and reduces injuries. Efficient production requires maximum use of modern equipment.

4. Providing self-watering devices.
5. Providing suitable protection.
6. Providing suitable storage.
7. Providing feeding equipment.
8. Providing loading chutes.
9. Providing proper restraining equipment.
10. Providing adequate scales.
11. Providing suitable wash areas.
12. Saving construction costs.
13. Handling waste materials.

1. Providing Proper Fencing

Fencing Essential—Proper fencing makes possible the application of many approved practices. Pasture

Fig. 6.1—This is not the way to load cattle. Proper equipment will save time and labor as well as avoid injuring animals.

rotation, deferred grazing, and a planned breeding program are possible if adequate fencing is provided. Ownership of cattle is seldom a problem, and no loss of time is encountered in constantly sorting cattle of various ownership. Good grazing within the area in which the cattle are feeding will put less strain on fences, as cattle seldom try to get out.

Select Proper Type—Fencing is expensive so it is necessary to select the type best suited for the operation. Cheap fence or fencing seldom is a sound investment in the long run. In large pastures or on the range, barbed wire fencing is very adequate. Four strands of barbed wire are most desirable, although three strands of wire are satisfactory on extremely long runs. In deep snow country, barbed wire fences should be con-

Fig. 6.2—An arrangement to show how cable fences are tightened and kept taut by use of turnbuckles.

structed so they can be taken down during the winter. However, wherever cattle are to be handled or "crowded" as in a corral, stout board fencing is desirable. If small calves are to be kept in or hogs are to follow with cattle, then woven wire fencing with several strands of barbed wire above should be used. Al-

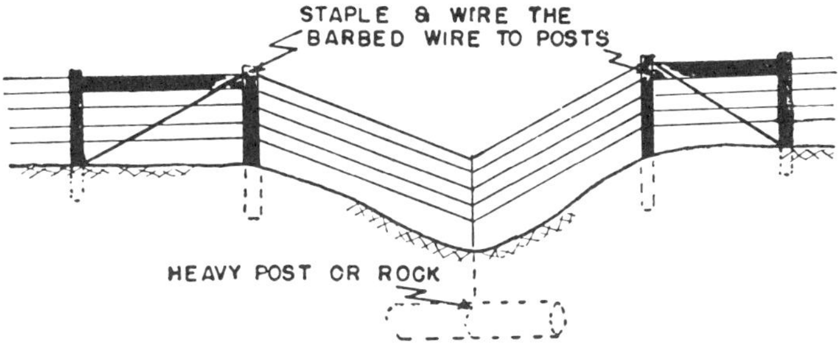

Fig. 6.3—Fences that cross low places can be "pulled down" through use of a deadman or rock buried in the lowest spot. Fasten a wire to the fence above and around the deadman or rock. This will hold the wires down in the low place. The post on either side of the low place should be braced like an end post, using a horizontal brace and diagonal wires. Where fences cross a deeper place, a separate fence should be constructed in it. Or, if water flows in it, self-cleaning floodgates should be used.

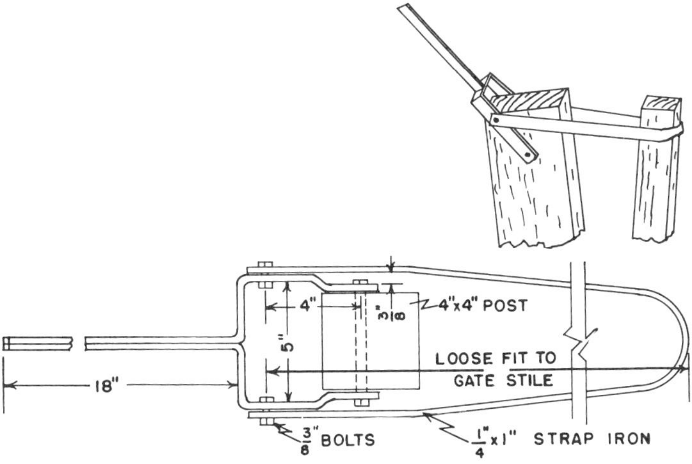

Fig. 6.4—Gate fastener. This gate fastener is simple to make and is very effective when used on wire gates. It provides a quick, easy way to keep gates tight.

ways use heavy type wire fencing as light gauge wire fencing continues to stretch and sag after construction.

Treat Posts—Posts should be chemically treated to prevent rot and decay regardless of where they are to be used. A large part of the cost of building a fence is labor; therefore only treated posts should be used. Posts most adaptable to the area should be used, as climatic conditions vary in different localities. Normally split posts outlast poles.

Space Posts Correctly—Spacing of posts will vary from 3 or 4 feet apart on sorting or handling corrals

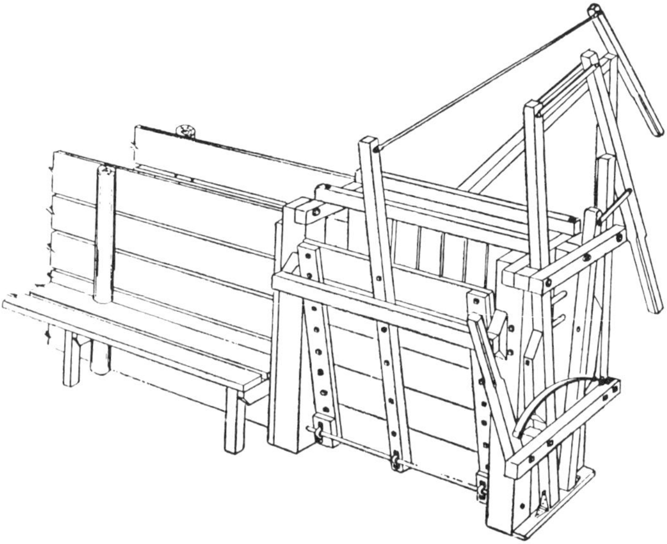

Fig. 6.5—A "squeeze" is the key piece of equipment for handling and inspecting cattle. The Nevada plan shown here has been used by many cattlemen and is considered one of the best designs. A number of very satisfactory all-metal prefabricated squeeze chutes are on the market. A squeeze chute plan is available through your local county extension office. The branding chute leading from the pens or yards to squeeze should be approximately 30 feet long to hold 4 or more head.

to 15 or 20 feet on the range. However, most livestock men seldom put posts more than 12 feet apart. Six or seven foot posts put 24 to 30 inches in the ground are desirable for most fences. Corral fences should be from 5 to 6 feet above ground and set from 30 to 36 inches underground. If cattle are especially difficult to handle, corral fences must be even higher.

The distance between posts determines the number required per mile—12 feet apart 440, 15 feet apart 352, and 20 feet apart 264 exclusive of corner posts. Sixteen rolls or spools of barbed wire are required to construct a mile of four-wire fence. One roll of barbed wire contains 1,320 feet.

Keep Bottom Wire High—The lowest wire should be high enough so very young calves can squeeze under. Sixteen inches from the ground seems to work

Fig. 6.6—Corner post properly braced.

well. Four wires properly spaced is better than five or six spaced wrong.

Brace Corner Posts—More strain is on the corner posts than any others which necessitates additional bracing. These posts should be set first in order to properly stretch and maintain tension on wire during construction.

It is a desirable practice to brace one line post every 200 or 300 feet so that fence remains true.

Use Boards for Corrals—Cattle can see board corrals and will not attempt to go through when being crowded. "Five quarter" lumber or 2x6 boards spaced 4 to 6 inches and bolted to posts form a desirable corral fence. Poles are a desirable item to use for corrals in areas where they can be easily obtained. Round off corners to prevent injuries.

Fig. 6.7—A handy stile makes it easy to step over an electric fence. Such devices are simple and easy to build and reduce the number of gates needed. They greatly facilitate the inspection of cattle by guests or prospective buyers.

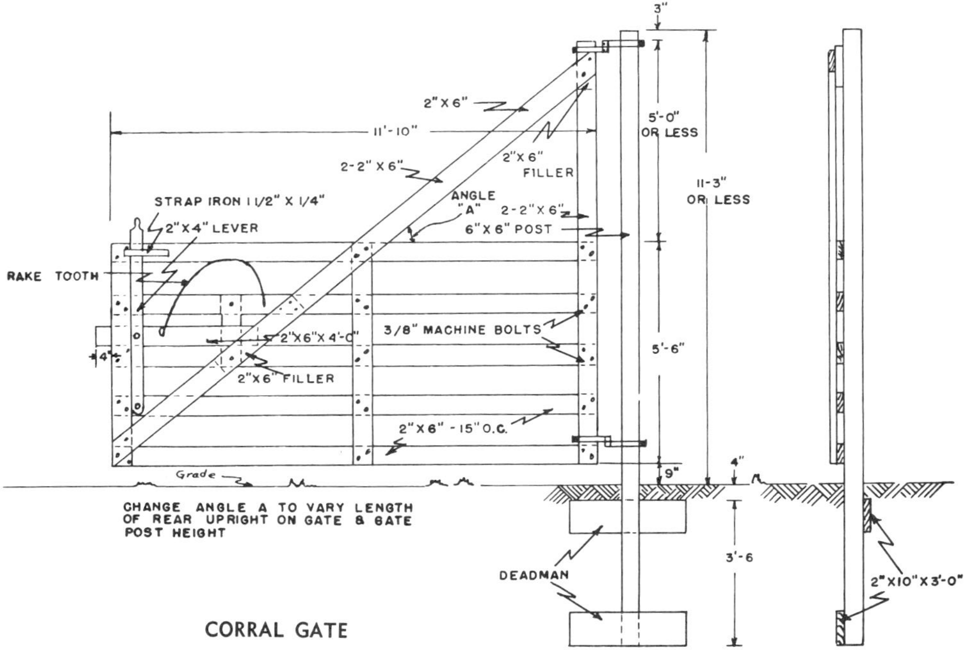

Fig. 6.8—Corral gate.

Using stock sizes and economical lengths and grades of lumber.

TYPE A.

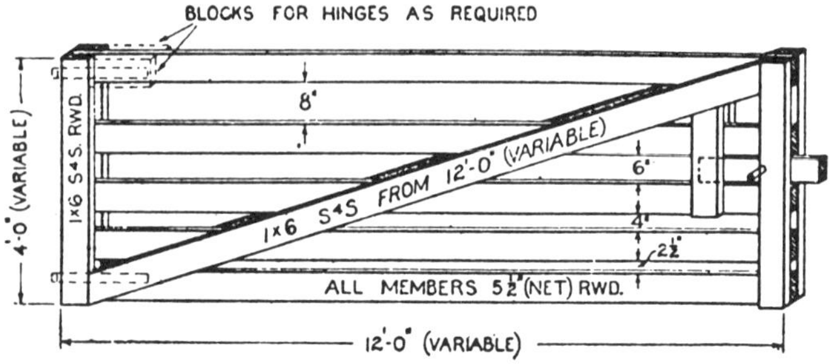

TYPE B.

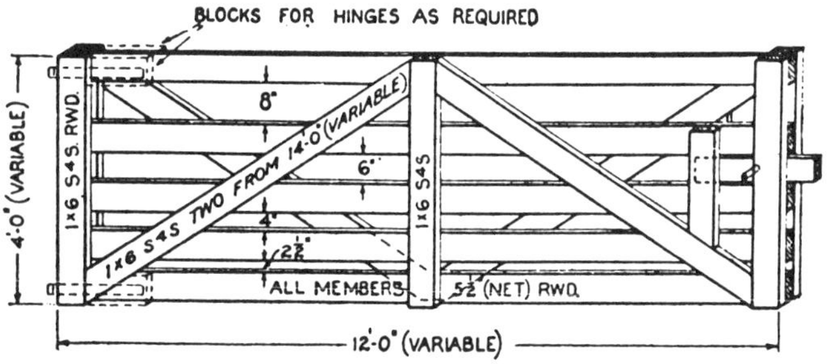

Courtesy, California Redwood Assoc.

Fig. 6.9

Use Electric Fences—For temporary fencing or to assist in management of irrigated pasture, electric fences will do a reasonable job at little initial cost. However, they should not be used for permanent fencing or wherever cattle will be crowded in small areas or for corrals. Electric fence should not be used as outside or boundary fences.

Fig. 6.10—A unique practical gate latch. It is constructed of one-inch pipe with a three-quarter-inch solid metal rod sliding inside. Gravity keeps the gate secured.

2. Providing Proper Gates and Cattle Guards

Provide Gates—Proper gates correctly placed on the cattle ranch insure a saving in time and labor.

Gates Should Be Well Constructed—Gates properly constructed last longer, give a better appearance, and will not sag. Bolts should be used throughout and strong lightweight lumber is better than extra heavy boards except for corral gates where cattle must be crowded. As a rule, gates should not be over 12 feet wide.

Use Heavier Gates for Corrals—Corral gates should be well constructed of 2x6 lumber bolted together. However, it is a desirable practice to keep them as short

Courtesy, University of California, College of Agriculture

Fig. 6.11—A "wheel" design for cattle handling corral. The triangular shaped hub is hinged so that an operator can shunt cattle to any pen from his position in the center.

as possible, generally not over 8 to 10 feet long. Some commercially sold gates are adequate for this purpose. Gates constructed of light weight metals are obtainable and are very satisfactory.

Mount Gates High—Gates should be mounted so they will clear the ground from 10 to 12 inches. Gates so mounted will not sag on the ground.

Provide Cattle Guards—A real saving in time and labor can be effected by the installation of strategically placed cattle guards. Tractors and trucks can pass over them, but cattle cannot. Farmer's Bulletin 1584 suggests the plan for a cattle guard shown in Fig. 6.12.

Many adequate commercial cattle guards are available. It is a desirable practice to build and install cattle guards so they can occasionally be cleaned out. Desirable materials for construction are concrete and steel.

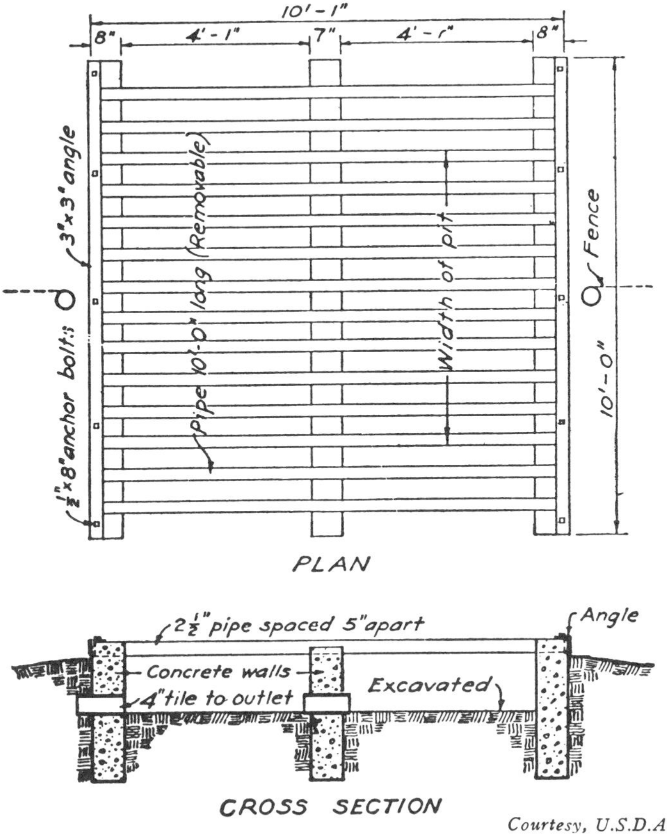

Courtesy, U.S.D.A.

Fig. 6.12—Plan and cross section of cattle guard.

3. Providing Proper Handling Corrals

Properly Locate Corrals—When cattle are handled in large numbers, corrals or pens should be located near a watering place. Time spent on planning, locating, and constructing pens is well worth-while because of the saving in man hours and the elimination of unnecessary handling of cattle. On general farms, many stockmen utilize one side of a barn or the area between

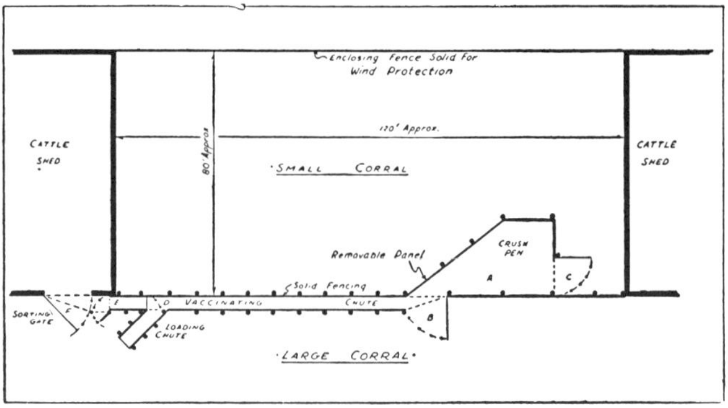

Courtesy, Country Gentleman *Magazine*

Fig. 6.13—Cutting corral. See page 221.

CATTLE CORRAL

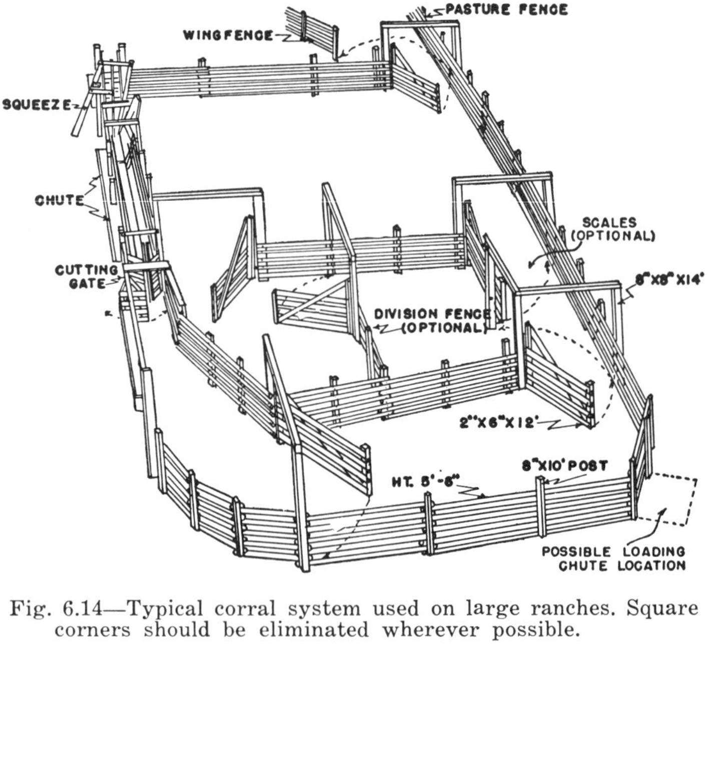

Fig. 6.14—Typical corral system used on large ranches. Square corners should be eliminated wherever possible.

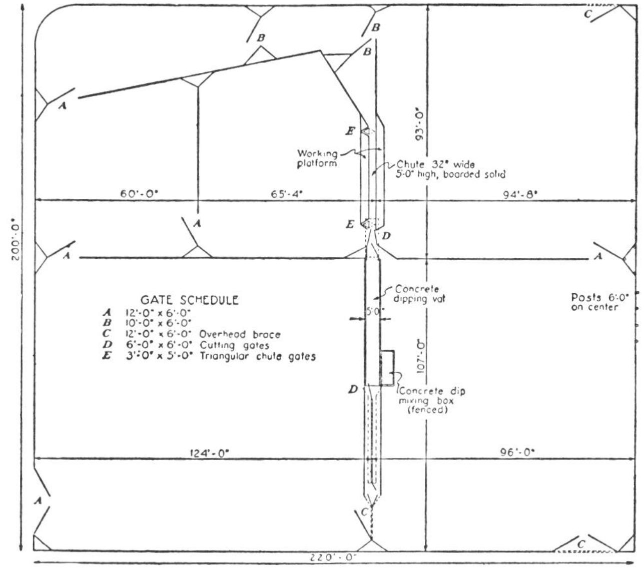

Courtesy, Country Gentleman *Magazine*

Fig. 6.15—A convenient corral system, showing arrangement of cattle-marking chute and dipping vat.

two barns as corral space. Corrals should be located so as to eliminate long drives in getting the cattle to them.

Properly Design Corrals—Even though the number of cattle to be handled is small, it is desirable to construct a system of corrals of two or more pens to permit "cutting" or separating cattle when desired. The following plan is suggested by Colo. State University where large numbers of stock are to be handled.

Provide Cutting Chutes—Cutting chutes with swinging gates used when separating various classes in separate pens, mean a saving in labor cost.

Gates should be well constructed and hung so as to operate easily.

Courtesy, John Bean Mfg. Co., San Jose, Calif.

Fig. 6.16—Cutting chute being used to spray cattle for cattle grubs.

Dipping Vats—Where it is necessary to dip large numbers of cattle, as in many sections of the range, country dipping vats are desirable. Plans can be obtained from local county agents. Spray-type portable dipping equipment is available commercially.

4. Providing Self-Watering Devices

Provide Permanent Tanks—Concrete tanks are used almost exclusively for permanent construction. Tanks are usually made round or square. Rectangular troughs permit more cattle to drink at one time. Concrete tanks should be reinforced.

Develop Springs and Reservoirs—Effective utilization of grazing areas is possible if adequate water is developed from springs and runoff. Proper development of water supply will make a greater amount of better quality water available for livestock. Most states have laws regulating the impounding of water and construction of dams, and these should be strictly followed. Soil Conservation Service can provide information regarding such items.

Regulate Water Automatically—An adequate supply

Fig. 6.17—A good type watering trough. Water is available on either side of the fence.

Courtesy, Soil Conservation Service, U.S.D.A.

Fig. 6.18—Recommended method for developing a spring to supply clean water.

of water with a minimum amount of labor involved is best provided by an automatic water regulating device. These are best obtained commercially. They should be protected so that stock cannot damage them or located so cattle cannot reach them.

Fig. 6.20 shows an automatic float and regulator installed in the center of a large trough so that cattle cannot reach it.

Provide Large Tanks—Water tanks should be large enough so that all cattle can drink without continually depleting it. Troughs should never be smaller than 200 or 300 gallons as water from large tanks keeps cooler.

5. Providing Suitable Protection

Keep Cattle Dry—Beef cattle will withstand extremely cold weather if kept dry. Cattle having access

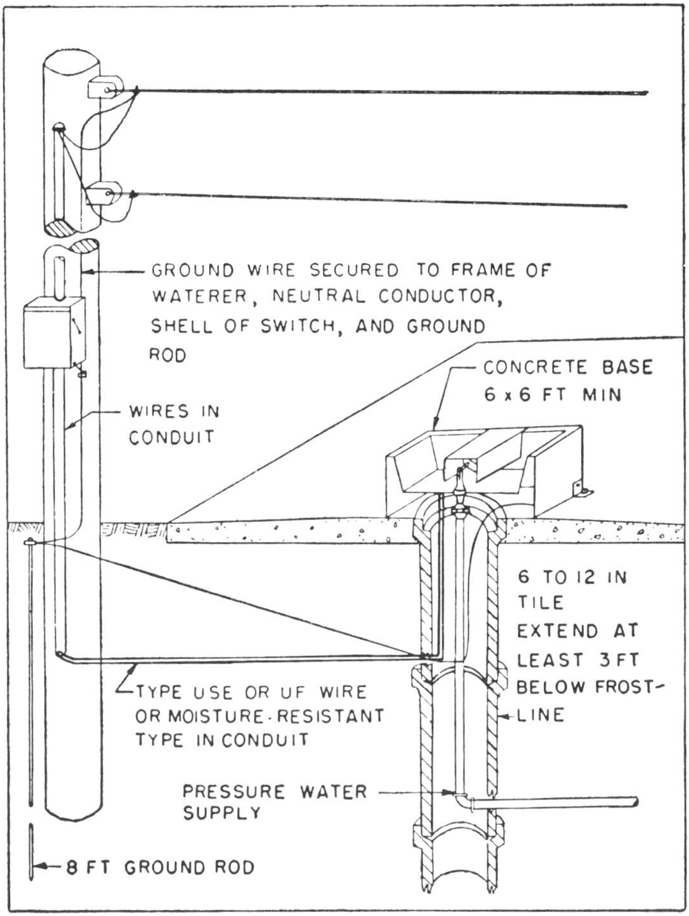

Fig. 6.19—Drawing showing installation details of typical automatic, electrically heated livestock waterer.

to bedding, both under cover and in the open, frequently show a preference for the latter. Especially is this true during weather which is extremely cold but dry.

Courtesy, Basalt Rock Co., Napa, Calif.

Fig. 6.20—Circular water tank. Diameter is large enough so cattle cannot reach float from any angle.

Provide Southern Exposure—Sheds with southern exposure provide adequate protection for beef cattle. The shed shown in Fig. 6.21 is suggested as protection utilizing southern exposure.

Provide Cattle Sheds—A shed 100 feet long and 20 feet wide accommodates 100 steers where they are unconfined or 50 steers if confined to shed. The masonry cattle shed shown in Fig. 6.22 is recommended.

Sheds can also be used for feeding pens by utilizing movable feed racks.

If cattle have access to yards, 20 to 25 square feet of shelter space is needed for each calf—30 square feet for yearlings, and up to 45 square feet for big cattle. Many cattle feeders crowd their cattle into as little as ⅔ of this space and with plenty of bedding, still get good results.

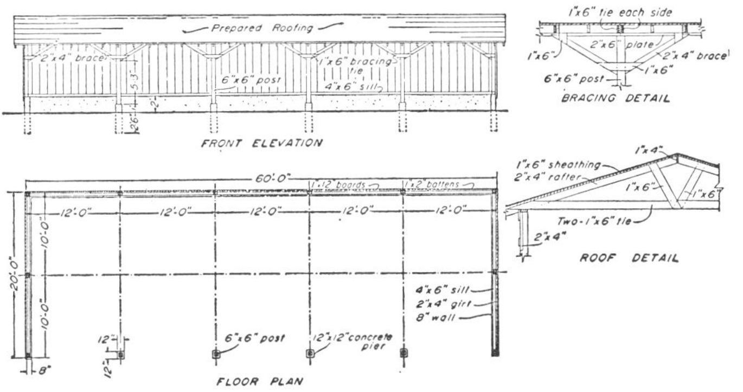

Fig. 6.21—Front elevation, floor plan, bracing, and roof detail of open shed.

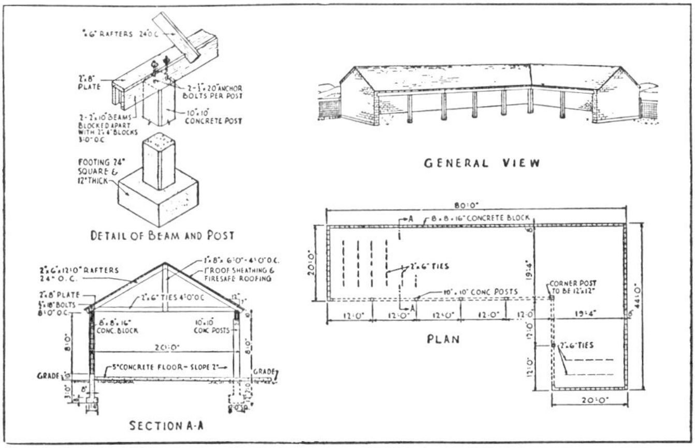

Courtesy, Portland Cement Assoc., Chicago, Ill.

Fig. 6.22.

Provide Windbreaks—They are especially desirable in pastures used for winter grazing where little or no natural protection is afforded. A suitable windbreak is a board wall 8 feet high. Windbreaks should be placed at right angles to direction of prevailing winds.

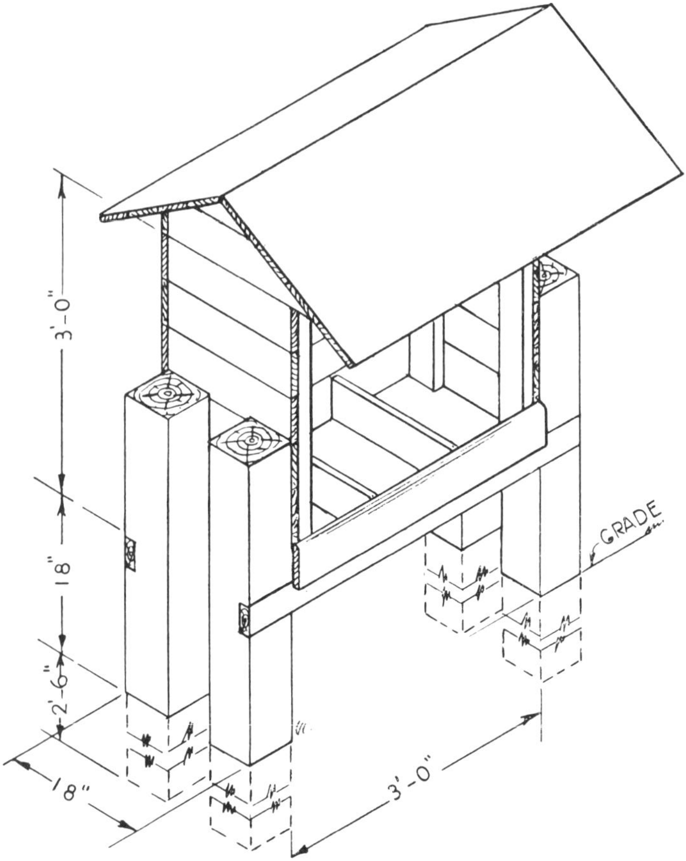

Fig. 6.23—Covered mineral or salt box.

Utilize Loft Storage—The repeated hauling of feed and bedding to the feed lot can be eliminated by utilization of loft storage. It is low cost storage. Treated poles can be utilized for both the frame and the foundation and will insure rigidity at low cost. This type of construction will make the shed serve as both shelter and storage.

Fig. 6.24—Pole-type open-side storage barn used to store hay.

Provide Only Necessary Protection—Cattle are able to withstand considerably more weather than commonly supposed if adequate food is supplied. For economic reasons, therefore, only shelter and protection should be supplied as the particular locality demands. In many localities which have a mild climate, protection may be unnecessary. Natural protection, such as trees, hills, etc., may provide adequate protection.

Provide Shade—Intense heat may affect gains as much or more than cold. Trees and green pastures provide the best temperature regulators. However, if summer temperatures are high and no natural shade is available, open-side shelters or "breezeways" should be provided.

6. Providing Suitable Storage

Have a Safe Margin—Storage capacity should always be such that the farm has on hand slightly more feed than is necessary to carry the herd through the winter. This is a desirable practice even if the feed in storage is only straw or some other kind of cheap roughage.

Provide Silos—A silo offers the best means of saving and storing the entire corn plant for feed. However, almost any crop can be made into silage if adequate preparations are made. Many satisfactory commercial silos are available, although farmers can construct their own at a considerable saving in labor cost.

Utilize Trench Silos—They can be built with little cash outlay and give fairly satisfactory service; however, spoilage is greater than in permanent type silos. Trench silos built in heavy soils are the most satisfactory. The capacity in tons of a trench silo can be calculated by multiplying the depth by the average width, multiplying this result by length, and dividing the answer by 60 (measurements all in feet).

Protect Hay—Protection against sun and rain is especially important in maintaining quality and nutritive value in hay. Dampened hay may become spoiled and moldy and even be dangerous to feed to cattle. Pole-type open-side storage barns are an economical way in which to store hay, especially if it is baled.

Combination storage and feeding barns are recommended in areas where severe winter climates prevail, especially for purebred herds where the cattle must be more closely watched.

There are many factors to consider in selecting a barn. The following suggestions may be helpful in selecting the type to build:

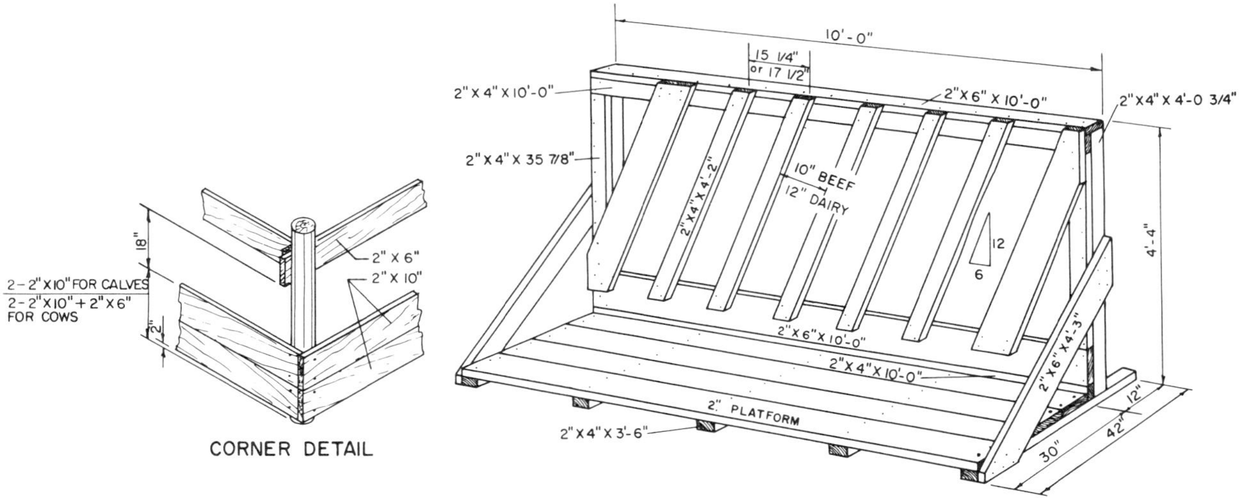

Fig. 6.25—Plan for hay and silage feeder.

a. It must protect feed as well as store it.
b. It should be designed to conserve labor and time.
c. It must be built well enough to withstand the elements of weather.
d. It should be designed to serve present needs and fit in with possible future changes in the farm program.
e. It should be economically constructed and in keeping with the productive capacity of the individual farm.

TYPICAL FLOOR AREAS FOR ANIMALS WITH ACCESS TO OUTSIDE YARDS

Animal	Conditions	Floor area (sq. ft.)
Breeding cow	With or without calf	50
Calf	Feeders, stockers, and replacement heifers	30
Yearling	Feeders, stockers, and replacement heifers	40
Fattening stock	Av. 750 pounds for fattening period	45
Fattening stock	Av. 950 pounds for fattening period	50
Bull	In pen	120
Cow	In maternity pen	100 to 120
Cow	In standing stall 4′ wide, 5′ to 6′ long, platform only	22
Calf	Several in pen, each	20

RECOMMENDED MINIMUM WIDTHS FOR SERVICE AREAS

Service passage	Conditions	Width
Feed alley	For feed cart	4′-0″
Driveway	For wagon, spreader, or truck	9′-0″
Doors and gate	Drive-through	9′-0″
Doors and gate	To small pens	4′-0″

RECOMMENDED DIMENSIONS FOR FEEDING DEVICES

Device	Conditions	Width
Feed bunk for grain or silage	Feed from both sides	3′-0″ to 3′-4″
	Feed from one side	2′-6″
Combined rack and bunk	Feed from both sides	5′-0″
	Feed from one side	2′-6″
Hay rack	At side of barn or pen	2′-6″

Trough height, 30″ for mature or fattening stock; 24″ for calves.
Trough length, 30″ for mature animals; 20″ for each calf.
Self-feeder trough length, 9″ per head (if feeder is kept filled).
Bunk or rack length for movable structures, 12′-0″ to 16′-0″.

Process Hay—One of the most costly items in providing roughage for supplementing or winter feeding beef cattle is that of handling hay. Any system that can cut down on the amount of labor and speed up hay handling should be fully investigated. Following are several suggestions that may prove helpful.

1. **Barn Dry Hay.** Hay can be stored with slightly more initial water content and with improvement in quality by barn drying. Most important, it can be brought in from the field quicker should rain threaten. The added costs of ducts and air blowers installed in the barn should be weighed against the advantages.

2. **Field Chop Hay.** Increased storage, elimination

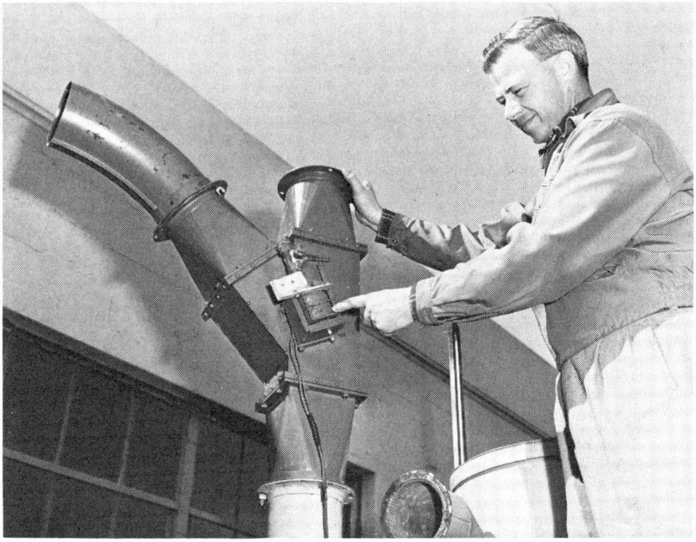

Fig. 6.26—The device shown eliminates foreign metal particles from hay or other forage going through the blower. Whenever foreign matter passes the sensing device it activates a gate to bypass the objects which could cause hardware disease.

Fig. 6.27—Chopped hay stored in temporary bin.

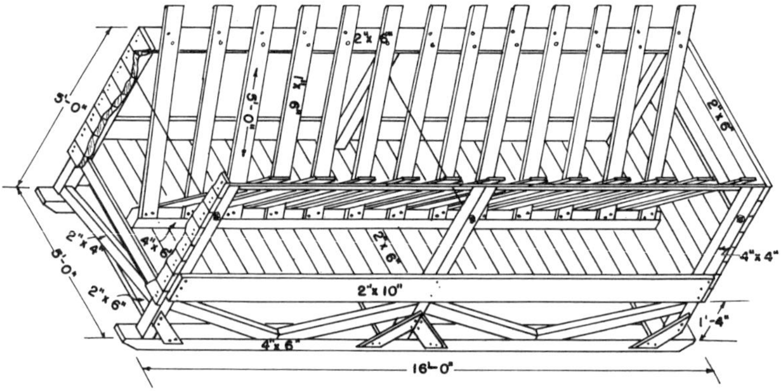

Fig. 6.28—Portable feed rack. There are many different types of hay feeders; one type often seen is shown here. This portable rack for feeding hay can be used in the corral or pasture. When filled, it will hold 160 cubic feet—even more, if it is rounded off at top. This rack is rugged in construction. It catches the leaves and fine stems that drop off as the hay is pulled out between the slats. That means a saving of the best part of the hay.

However, this type feeder is sometimes criticized as wasting too much hay. But correct spacing of the slats will solve the wastage problem. For flaked baled hay and long hay, slats should be from 3½ to 5½ inches apart. If round bales are to be fed, the spacing should be about 7½ inches.

of almost all hand labor, and an improved product are often the result of mechanical field choppers. It is possible to store chopped hay in temporary silos made of fencing and protected on top from rain. Chopping works especially well with uniform roughage, such as alfalfa. Advantages should be balanced against the cost of mechanization. Local county agents can advise on matters of chopping hay or barn drying for a given locality.

7. Providing Feeding Equipment

Feeding equipment should be heavy enough to withstand abuse but light enough to handle. If possible, equipment should be on skids so it can be moved from place to place.

Utilize Hay Racks—A well constructed portable hay rack with trough bottom will save hay. Roughage on the ground is trampled and wasted.

Make Silage Bunk—A good silage bunk allows animals to feed from both sides without interference with one another, yet narrow enough so cattle can reach all the silage.

This bunk could also be used for a feed trough.

Provide Self-Feeders—Most cattle are hand fed. However, once the cattle are on full feed, very little difficulty should be experienced with the self-feeder. Plans for a portable self-feeder are shown in Fig. 6.28.

8. Providing Loading Chutes

All cattle ranches, even small ranches, should provide a chute in order to load and unload cattle safely and rapidly.

Properly Locate Chutes—Loading chutes must be

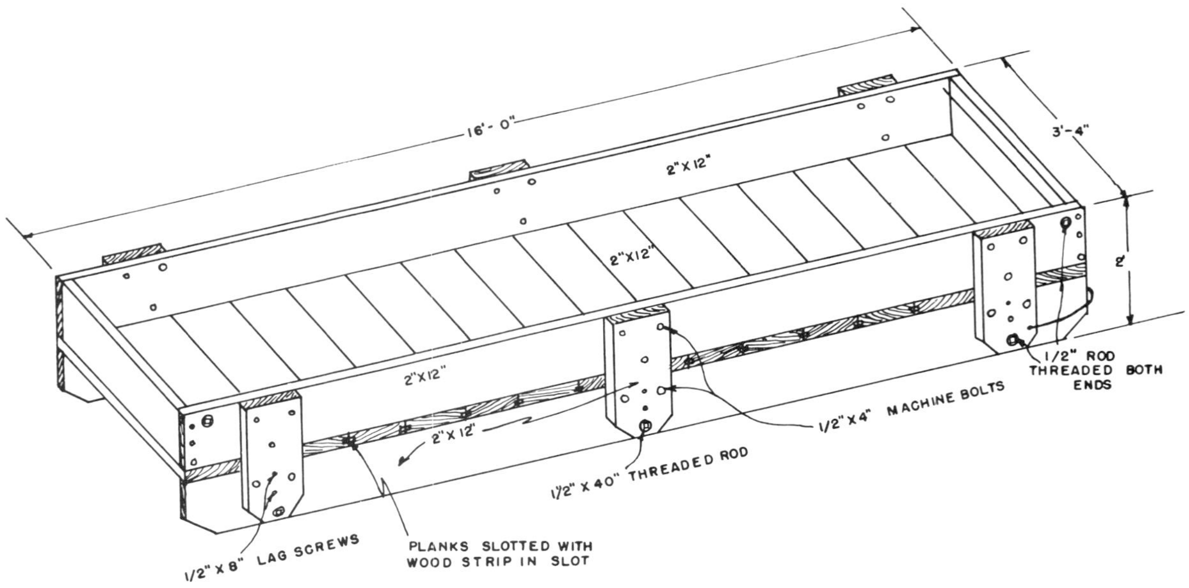

Fig. 6.29—Portable feed bunk. This bunk has been used by feeders and has stood up against very rugged use. This bottom of the bunk is made of treated planks. The planks are slotted and, as they are laid, a special wood strip is placed in the slot. This prevents cracks from forming between the bottom planks, eliminating grain and feed loss.

The side uprights are fastened to the sides and to the bottom runners with lag screws. A 2″ x 6″ is set crosswise under the bunk bottom between the uprights. Rods half an inch in diameter and threaded on each end are used as noted in the plan.

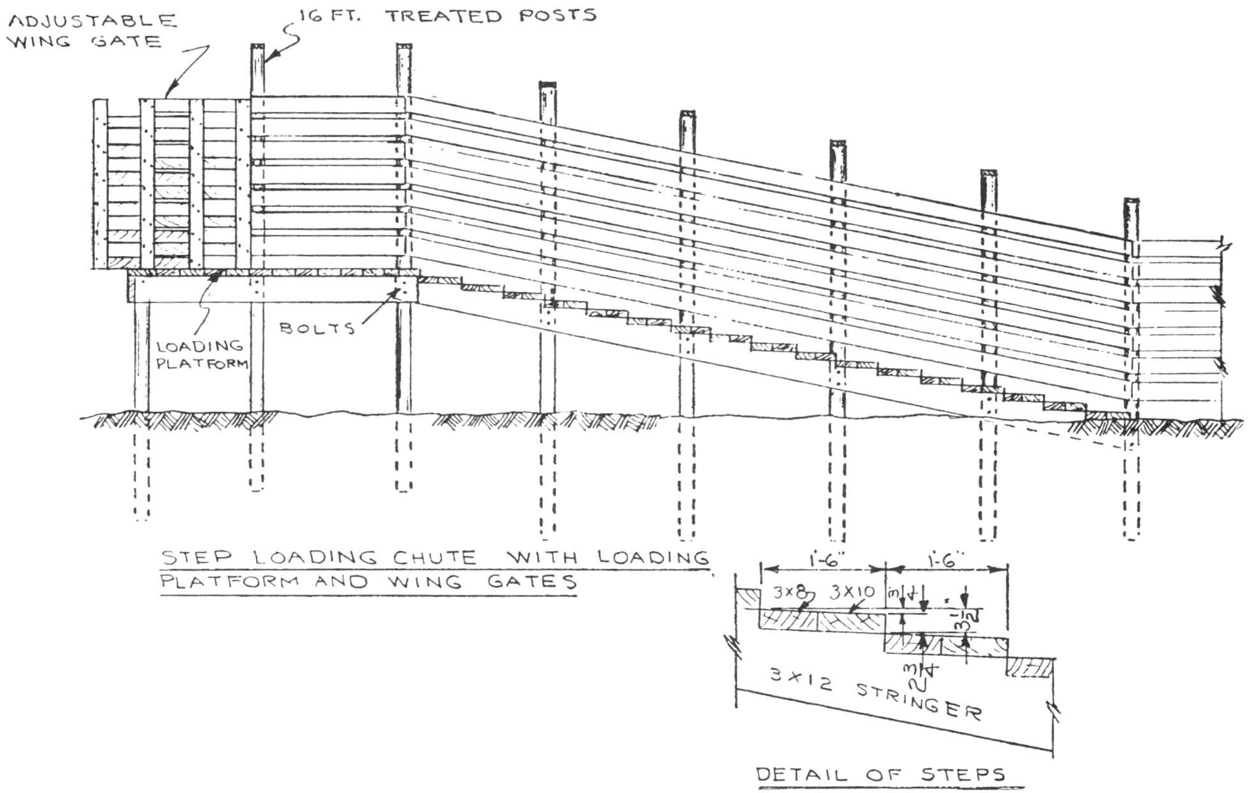

Fig. 6.30—Cattle will walk up step ramp loading chutes with less slipping and consequently less injury than on cleated inclines.

located near roads or in areas where vehicles can approach them safely in any weather. There must be adequate turning room for large vehicles. It is desirable to locate chutes near scales and holding corrals.

Properly Construct Chutes—Loading chutes should be stoutly constructed of surfaced heavy boards bolted together. If cattle are extremely wild, pens and alleys leading to chutes are sometimes 8 feet high. However, most cattle can be handled in chutes several feet lower. Standard height of floor for chutes designed to load to trucks is 42 to 48 inches from the ground. Most important factor in getting animals into chute is that it be solidly constructed. Animals are afraid of insecure footing.

Utilize Portable Chutes—Small farms often can utilize an all-purpose chute. It can easily be moved from place to place by mounting it on skids or a pair of wheels. Height of this type of chute is easily adjustable.

(Plan No FE 27)

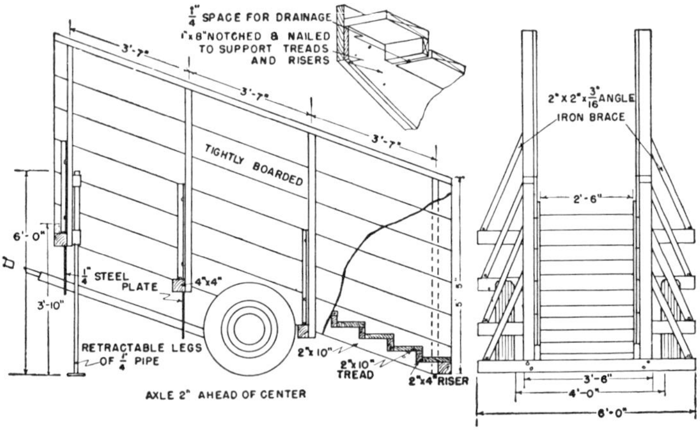

Fig. 6.31—Portable loading chute (stair-step type).

Fig. 6.32—Cattle loading chutes on beef unit.

Put a Curve in Chutes—Cattle tend to enter and go through a chute with less balking and greater speed if they cannot see through to the end. Many designs are used to create this effect but a simple and effective way is to curve the chute into a crescent shape or include a right angle or corner in its design rather than have it straight. Chutes built this way are stronger

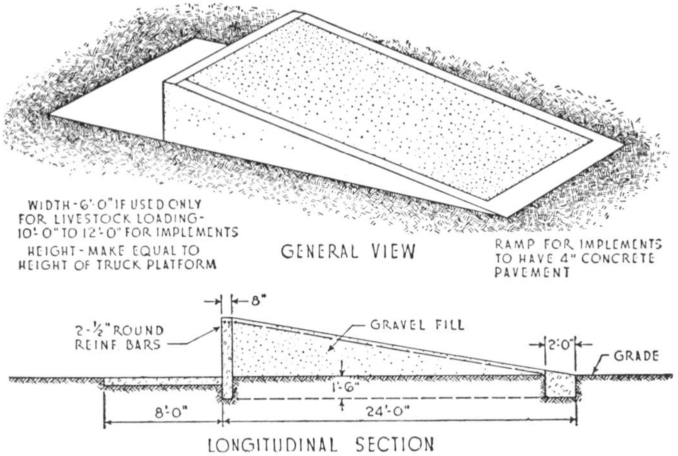

Courtesy, Portland Cement Assoc., Chicago, Ill.

Fig. 6.33—Concrete loading ramp.

Courtesy, Teco Products, Madera, Calif.

Fig. 6.34—Approved type calf squeeze.

Fig. 6.35—Cattle stocks utilized for hoof trimming. Note post in front needed to secure head of cattle.

and often lend themselves to a more usable kind of corral.

9. Providing Proper Restraining Equipment

Adequate restraining equipment is essential for success in handling cattle on modern farms.

Provide a Squeeze—Castration, branding, vaccination, and many other operations can be safely and quickly performed with a desirable type cattle squeeze.

Provide Stocks—This device is especially desirable in purebred herds where cattle are kept around corrals continually. The weight of the animal is taken off its feet so the feet can be trimmed or other operations performed.

Desirable plans for constructing an all-steel stock

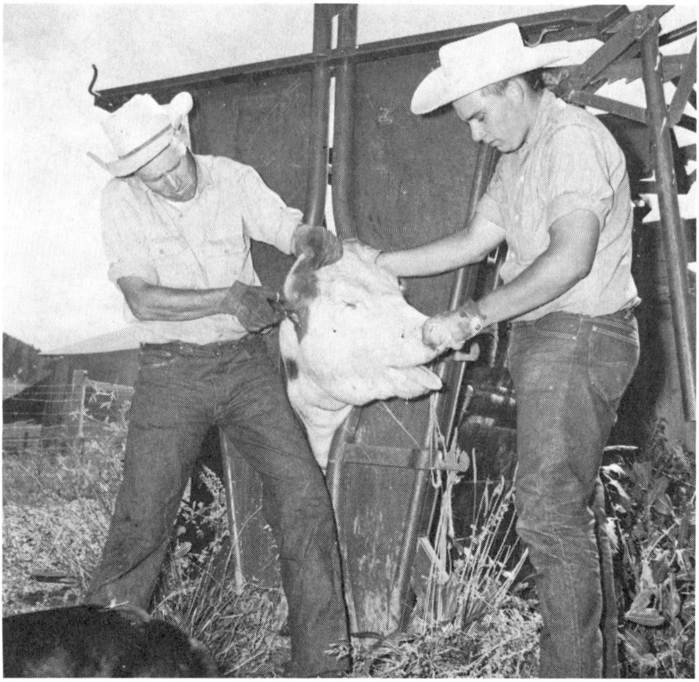

Fig. 6.36—Combination head gate and cattle squeeze.

can be obtained from Extension Service, Kansas State University, Plan No. 38.

Utilize Headgates—Dehorning, tattooing and vaccination can best be accomplished on a swing-out type headgate. Cattle are held securely enabling the operator to perform the task comparatively easily. Many desirable commercial headgates are available, although farmers can build satisfactory ones themselves if they desire.

10. Providing Adequate Scales

Weighing scales provide an accurate rapid method

Fig. 6.37—Scales and pens arranged so as to facilitate **easy** handling of cattle.

of weighing cattle and stimulate sales of cattle. They prevent sale of livestock "by the head" with results often detrimental to the producer. In addition, rate of gain can be checked frequently to assist in feeding operations. Scales should be located conveniently close to corrals and loading chutes with ample runways to facilitate handling.

11. Providing Suitable Wash Areas

Purebred producers who have occasion to fit cattle frequently should provide a suitable wash area. Concrete flooring with an ample drain to carry the water away is desirable. The rack should be located out of the wind and open to the sun in order to keep the animals as comfortable as possible. A stout post or ring should be provided to secure animals.

Fig. 6.38—A portable scale that can be moved easily from place to place. Adequate scales are a must in any up-to-date system of breeding and improving livestock. Periodic weighing of random samples of cattle can indicate the progress of the entire herd or feeding operation.

12. Saving Construction Costs

Take Advantage of Terrain—Use of the slope of the land or existing windbreaks can assist in lowering construction costs. Loading chutes, for example, that take advantage of a hill or rock shelf can be built to facilitate loading at little expense.

Use Native Materials—Steel fence posts may be the best buy in many areas; however, in others split posts made on the farm, or even pole or log corrals, may be best if they are easily obtainable. Rock fences can be used at times although present day labor costs generally prohibit their use.

Build in Season—Post holes can be easily dug, land

Courtesy, Daniel Cassard

Fig. 6.39—A beautiful set of stone corrals. This is on an Alpaca breeding station in Peru where no trees are near for many miles, hence the use of native materials. It is high so it would hold sheep or cattle although in this instance it is used for llamas.

Fig. 6.40—A properly constructed wash area is needed in order to fit cattle. Floor is concrete and located on sunny side of barn and out of the wind.

Ring gate latch for gate. Latch is welded to gate post. Pin on gate lifts ring when gate is closed, but cannot lift it for opening. This latch can be made in a small shop if welding equipment is available.

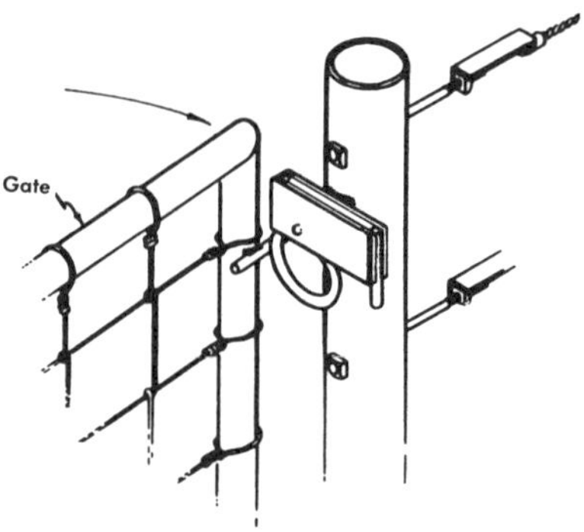

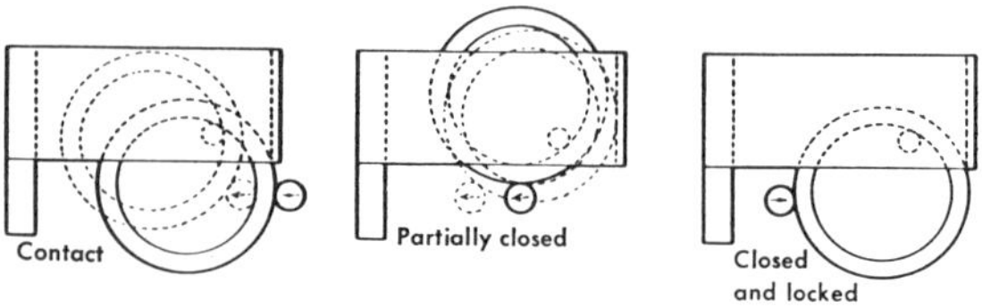

Fig. 6.41—Ring gate latch.

leveled or rocks removed when soil is damp from spring rains. Similarly, advance planning will allow building after rainy season is over so labor costs are lower as materials do not have to be put away each night or covered from rain.

13. Handling Waste Materials

In recent years odors from feedlots, buildup of manure, fly problems, and related items have become a concern of the livestock industry. While not so great a problem with beef cattle as with dairy, swine, or poultry, it is an issue especially around feedlots near populated areas. Manure is of little value unless spread on growing crops or pastures. While this waste material has a value, often the cost of removal is as much as or greater than its value as a fertilizer. For this reason many systems have been devised in order to lower the cost of getting rid of waste material.

Build Facilities for Power Equipment

Pens, corrals, alleys, and so on should be wide enough to allow for the use of power equipment to assist in clean-up. Corrals and feedlots should be large enough to permit easy access and turn-around space for power equipment.

Pave Lots

Concrete or blacktop in areas that get lots of use will prevent mudholes and permit use of power scrapers for quick and easy waste material removal.

Use Topography to Advantage

Building on well drained land or on a slope will facilitate drainage and eliminate some problems. Location in relation to prevailing breezes may prevent odors from being so noticeable and cut down on undesirable publicity.

Design for Automatic Handling

As labor costs increase, more attention will be given to mechanical disposal of manure. This would be more applicable in large feedlots where numbers of cattle and feed are concentrated. The use of slatted floors and automatic cleaners utilizing water and mechancial sweepers will tend to keep pens dry and livestock clean and at the same time minimize actual hand labor. If the waste material can be processed and sold as a by-product, a profit might be realized or at least the cost of operation reduced.

Fig. 6.42—Slatted floors used in feed lot. The slats are made of reinforced concrete spaced about 2 inches apart, so manure falls through and slats stay dry at all times.

Keep Ahead of Build-up

If manure is a problem, letting it get out of hand compounds the situation. Every effort should be made to prevent excessive build-up with resultant fly problems, odors, packing, unsightliness, and so on. Frequently, existing equipment will not handle overloads. Insufficient water pressure and volume will not wash away manure properly. However, a regular systematic clean-up lets all systems function as they should.

Fig. 6.43—This photo illustrates the manner in which waste materials—manure, etc.,—are removed from space beneath the slatted floor. Water plus a drag pulled by a cable continually removes sludge.

Chapter

VII

CONTROLLING PARASITES AND DISEASES

Control Measures Essential—One of the more important limiting factors in beef production is the loss from parasites and diseases. Most losses in beef production can be prevented if approved practices are followed in combating parasites and diseases. Consequently, it is imperative that cattlemen know what approved practices should be used and formulate plans for carrying these practices to successful completion.

Activities Which Involve Approved Practices

1. Providing a satisfactory prevention and sanitation program.
2. Preventing bloat.
3. Controlling cattle grub.
4. Controlling screwworms.
5. Controlling external parasites.
6. Controlling internal parasites.
7. Controlling anaplasmosis.
8. Controlling anthrax.
9. Controlling Bang's disease.
10. Controlling blackleg.
11. Controlling choke.

12. Controlling foot rot.
13. Controlling IBR.
14. Controlling leptospirosis.
15. Controlling pinkeye (infectious keratitis).
16. Controlling red water disease.
17. Controlling scours.
18. Controlling shipping fever.
19. Controlling tetanus.
20. Controlling trichomoniasis.
21. Controlling vibriosis.
22. Controlling warts.
23. Controlling other serious diseases.
24. Treating wounds.

1. Providing a Satisfactory Prevention and Sanitation Program

Practice Sanitation—The importance of cleanliness in the care of animals cannot be overemphasized. Proper cleaning removes most of the germs along with the dirt that protects them, and the remaining germs are usually so weakened or so few in number that they are harmless under ordinary conditions. The ability for germs to invade the animal body depends primarily on the number present and their vigor. When needed, the proper disinfectant properly applied will go a long way toward eliminating danger from germs that cause disease.

Provide Safe Pastures—Beef cattle, as a rule, do not present too great a problem from the disease standpoint. One reason is because cattle are separated on large pastures and not crowded together as are many other kinds of livestock. Pastures should be well drained and free from bogs if possible. Any area that is dangerous because of poisonous plants, toxic conditions of the soil, or danger from disease germs

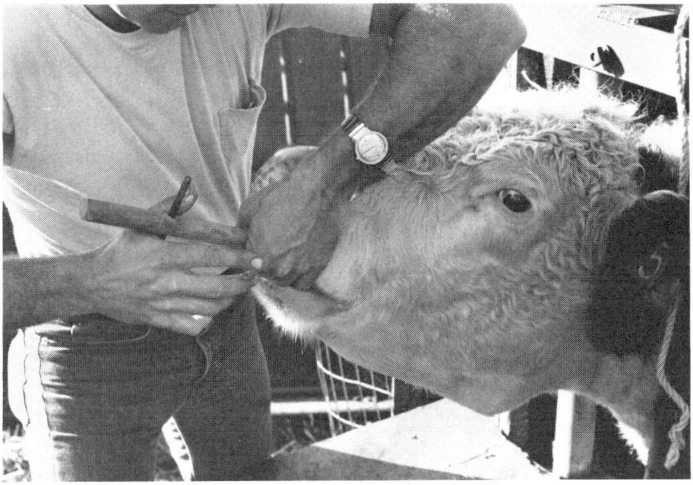

Fig. 7.0—Dispensing a pill or bolus is a skill all livestock persons should master. The side of the mouth of the animal is grasped by one hand in order to open the mouth and the balling gun pushed back on top of the back portion of the tongue. The pill must be placed far enough back in the throat so the animal cannot cough it up.

should be corrected or fenced off. Special attention should be given to drainage in irrigated pastures as internal parasites are especially numerous on pastures that are damp year round.

Eliminate Hazards—Old wire, nails in corrals, deep holes, gates in need of repair, and numerous other items around the farm are a constant source of injury and common cause of death to cattle. It is a desirable practice to eliminate such danger zones on beef ranches.

Keep Animals Healthy—Properly fed animals are not so apt to eat poisonous plants or have unusual cravings. If fields are not overpastured, cattle do not have to graze as close to the ground where they pick up nails and wire which may cause death. Farmer's Bulletin

Fig. 7.1—One method of taking a blood sample. The needle is inserted under the tail alongside the vertebra; some persons prefer to take blood samples from the top of the tail, as it "spooks" the animals less. Many diseases and conditions of animals can be detected via blood. All livestock persons should learn to perform this skill.

1991 states the healthy animal body is able to resist invasion by considerable numbers of germs.

Destroy Dead Carcasses—All dead animals regardless of the cause of death should be buried deep enough so dogs or other predatory animals cannot dig them up. Local reduction or tallow works will haul away, for a fee, large fresh carcasses like cattle or horses. Many zoos also will take carcasses free if they are fresh.

2. Preventing Bloat

Bloat, the overdistention of the first two stomachs with gas, is a serious problem in cattle production. It

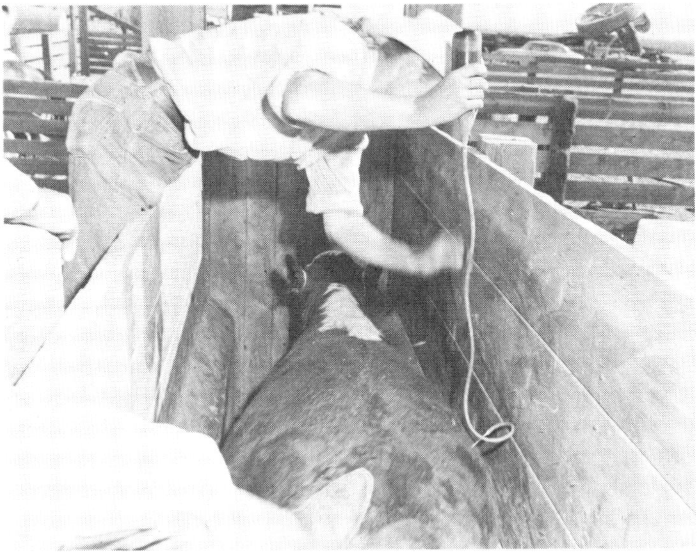

Fig. 7.2—Proper handling facilities are necessary in order to treat sick animals.

offers a real obstacle in the pasturing of legumes, such as clover and alfalfa. Bloat occurs on several other succulent feeds including roots and tubers. Occasionally an individual cow will bloat on any type of ration.

Present theory is that bloat is caused by any material (saponins) that will alter surface tension so that the gas of fermentation is trapped in countless bubbles throughout the feed mass in the rumen.

University of Wisconsin scientists have isolated an enzyme (PME) which is responsible for cattle bloating. It is found in large amounts in alfalfa and other legumes.

Supply Coarse Roughage—Bloat may increase from a dietary lack of sufficient fiber of proper type to initiate belching. Therefore dry hay, straw, or other types of coarse, stemmy feeds should be available to cattle at all times, especially if they are on succulent legume pasture.

Plant Grasses with Legumes—It is a desirable practice to plant grasses with legumes in a pasture. Bloat is seldom a problem if pastures are at least 50% grasses.

Withhold Hungry Cattle from Pasture—It is a desirable practice to feed cattle hay before turning them onto pasture so they will not be too greedy and gorge themselves on succulent feed.

Signs and Symptoms—Cattle and sheep will stop eating when they become bloated. As the bloating grows more marked, they become restless. If severely bloated, an animal is in great distress. It urinates and defecates frequently and lies down. Death can follow if the animal is not relieved.

Keep Animals on Their Feet—Should animals become severely bloated, every attempt should be made to prevent their lying down. Animals should be kept moving slightly until the excess gas is expelled by belching. A home remedy for arresting bloat is two ounces of turpentine and a quart of raw linseed oil. Never use boiled linseed oil. Another simple remedy for moderate cases is two tablespoons of turpentine or other antiferment in one quart of warm milk or water as a drench.

The *Yearbook of Agriculture* reports that defoaming agents like methyl silicone or peanut oil were effective in relieving and preventing bloat. However, it cautions that all bloat may not be due to frothing so that these agents would not be successful in all cases.

Call a Veterinarian—In a severe case of bloat, puncturing the paunch with a trocar may be necessary. This

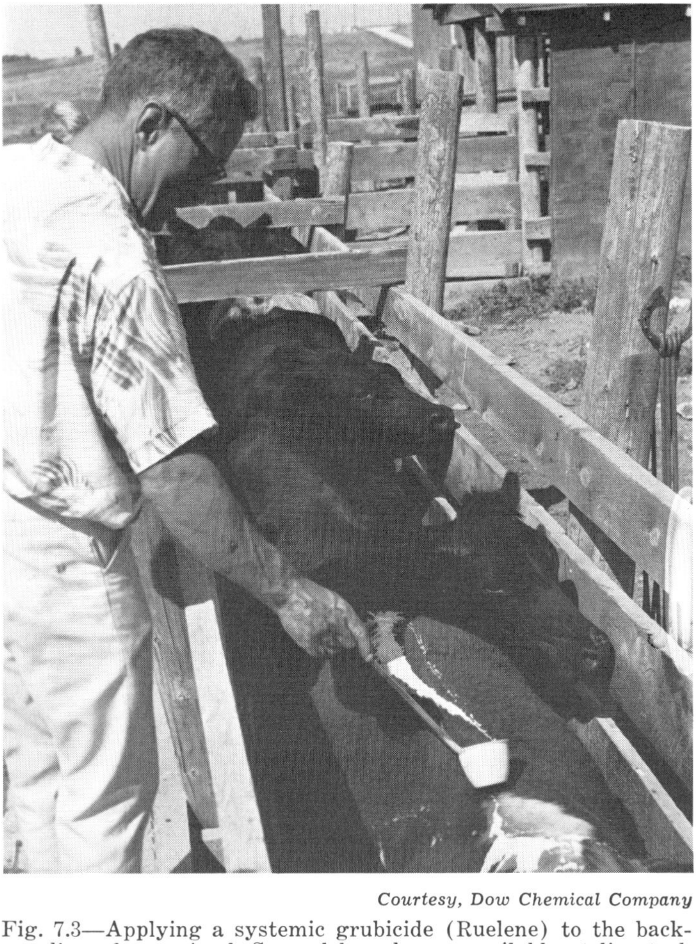

Courtesy, Dow Chemical Company

Fig. 7.3—Applying a systemic grubicide (Ruelene) to the backline of an animal. Several brands are available at livestock supply houses.

operation may injure the animal critically if not done properly, and it should be performed by a qualified veterinarian whenever possible. However, as beef ranches are often far from a veterinarian and time is impor-

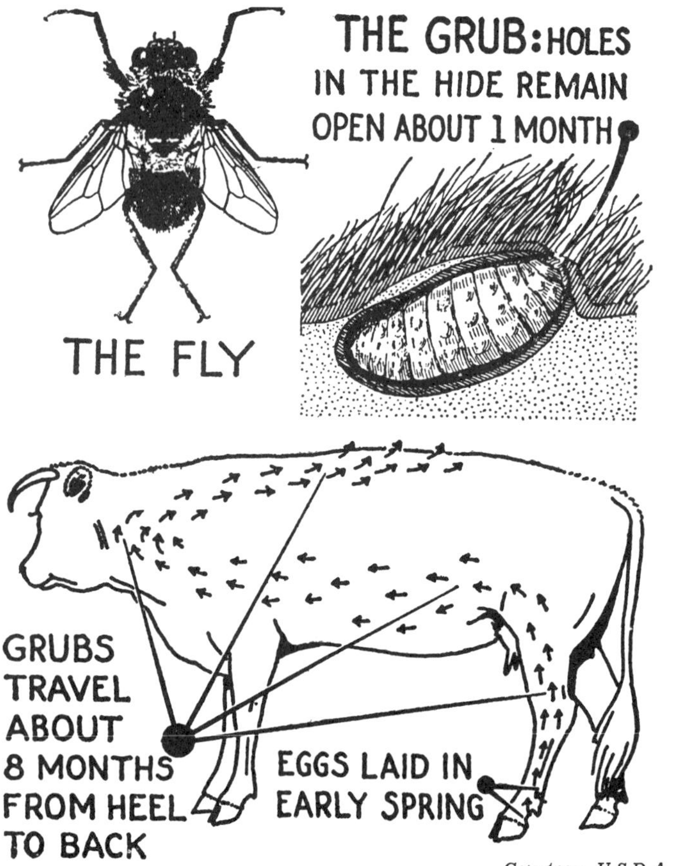

Courtesy, U.S.D.A.

Fig. 7.4—To kill grubs, the first application should be made just before they begin to drop from the animal's back. Time of year varies with locality. Treatment should be repeated at 30-day intervals. Usually three treatments are sufficient. Systemic materials are also used.

tant, it is often done by a rancher. In any case, puncturing the paunch should be a last resort.

The site to use is the paralumbar fossa or one-half way from hip bone to last long rib and 4 inches down from the short rib on left side.

Courtesy, Teco Products, Madera, Calif.

Fig. 7.5—Cattle being treated for grubs with portable-type spray dip.

If time isn't too critcial, call a veterinarian. He has more refined ways of relieving bloat and leaves no wound.

3. Controlling Cattle Grub

Importance of Cattle Grubs—Losses caused by the heel fly and its larva, the cattle grub, are estimated to be at least $50 million annually. The activity of one heel fly is at time sufficient to stampede an entire herd of placid cattle. Enumerated, the losses are caused by:

a. Tons of grubby meat which must be trimmed from highest priced cuts.
b. Leather lost because of grub holes.

c. Annoyance which causes cattle to run in panic and refuse to feed with resultant loss of flesh and injury from accidents.

Two species of heel flies parasitize cattle in North America. Both are much larger than the common house fly.

Cause of Cattle Grub—Grubs are the larvae resulting from eggs laid by the adult heel fly on the legs of cattle. Eggs are laid in the spring and the young larvae take about eight months to travel to the backs of cattle where they pierce the hide and mature. The complete life cycle is diagrammed in this chapter.

Symptoms—The most obvious sign of heel flies is in the spring when cattle run panic-stricken with their tails in the air, or hide in the brush and refuse to eat. Oftentimes they will stand knee deep in a water hole all day. Early in the year, the grubs can be felt as lumps along the backs of cattle when the area is rubbed with a hand.

Control of Heel Fly—Rotenone when available is a good insecticide recommended for grub control. The best time to apply the control is when grubs are in the backs of cattle. Rotenone may be applied by dusting, spraying, or dipping. Dusting is recommended for small herds. Mix one pound of 5% rotenone powder with 2 pounds dusting sulfur and apply to the backs of cattle with a large shaker. Rub the powder into the hair coat with the fingers. One pound of mix will treat 4 or 5 cattle. Commercial powders are available for use with small herds.

Spraying will give best results with large herds. Hold the nozzle 12 to 16 inches from the back and direct the spray straight down on the grub area. The first application should be made just before grubs begin to drop from the animal's back. Time of year varies with the locality. Repeat treatments at 30-day intervals. Usually 3 treatments are sufficient.

Use Pour-on Insecticide—See Fig. 7.3. This is probably the most common and effective way to control grubs and lice. Most ranchers use this method as it is easy and effective. Applied along the backline of an animal according to recommended dosage right after heel-fly season will control grubs all year.

Try Community Action—As the flight of the heel fly is quite limited, community action which involves control on all ranches within a given area may prove valuable. The area should be as large as possible and nearly square or circular in shape. An area involving several townships is recommended.

4. Controlling Screwworms

Importance of Screwworms—The insect pests are especially dangerous because the maggots feed on living flesh. The insect destroys untold numbers of domestic and wild animals annually not to mention the amount of labor and expense in treating infestations.

Cause of Screwworms—A fly about the size of the house fly lays eggs in irregular masses along the edges of wounds or blood spots of animals. These eggs hatch into small maggots in 8 to 12 hours. The maggots penetrate the area causing bleeding and irritation. As maggots destroy tissues, more flies are attracted to lay eggs which hatch into maggots. They are especially numerous in mild climates.

Control Measures—The following preventative measures are suggested for controlling screwworms:

a. If a controlled breeding plan is feasible, have

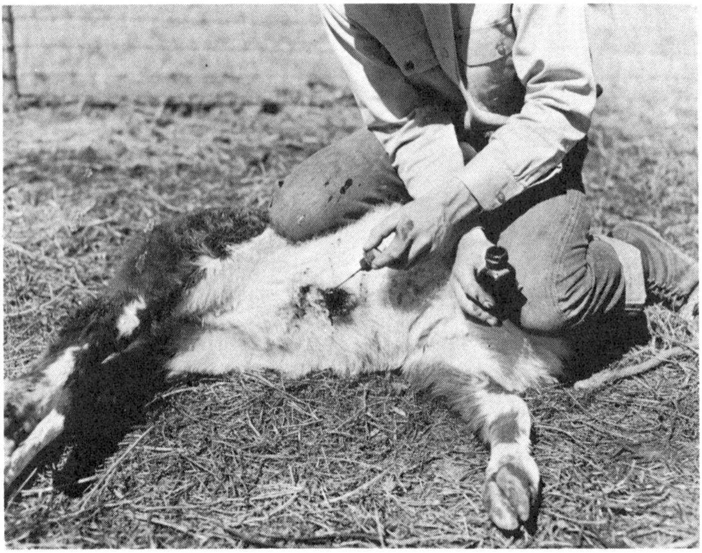

Fig. 7.6—Treating navel of calf infested by screwworms.

the calves dropped when there is little danger from screwworms.

b. Destroy all dead animals, preferably by burning.

c. Control external parasites, such as lice or ticks.

d. Avoid marking, castrating, and branding during screwworm season. Pincher type emasculators are helpful during screwworm season.

e. Apply U.S.D.A. Smear EQ335 which can be commercially obtained, to all wounds. The remedy is best applied with a small paint brush to infested or uninfested wounds. Two applications each week until the injured tissues have healed are usually sufficient to protect wounds against infestation. In many cases one treatment is sufficient.

Courtesy, Cutter Laboratories, Berkeley, Calif.

Fig. 7.7—The larva of the screwworm fly. This tiny maggot is one of the greatest enemies of livestock owners.

A new method is being tested as a means of controlling screwworms. It consists of releasing laboratory sterilized males which mate with native females who mate only once. Thus under certain conditions the pest could be completely eradicated.

This method has great promise, as the adult flies do not travel great distances, perhaps only a mile or two. Therefore entire areas could be made screwworm-free by the use of sterilized males. Cooperative action is necessary, but the rewards are great.

5. Controlling External Parasites

Importance—Flies, ticks, lice, and other external parasites cause tremendous loss to the livestock industry in a variety of ways. Animals rubbing on fences can injure themselves, making possible screwworm infestation. The annoyance of flies and lice causes

Courtesy, Old Scratch, Inc.

Fig. 7.8—Backrubbers are one method of applying insecticides to control and prevent external parasites.

cattle to become unthrifty so they do not gain weight or even maintain their weight. Ticks, for example, can transmit disease and until the end of the 19th century were perhaps the worst enemy of cattle raisers.

Symptoms—Cattle become thin and weak and are obviously annoyed. Often they fail to graze during daylight hours and seek waterholes, etc., in order to gain some respite.

Preventative Measure—Removing straw and litter from the barn at weekly intervals is most important in reducing stable fly population. The most effective control of flies is sanitation.

Use fly traps. Various types of fly traps are effective in reducing the population of flies around buildings.

Courtesy, Shell Chemical Company

Fig. 7.9—The fogging machine will effectively spray large areas to control flies.

The U.S. Bureau of Entomology has developed a fly trap for horn flies on range cattle that is very effective. It consists of a shelter 7 x 10 x 6 feet that has a series of Z-shaped screens that trap flies as cattle pass to and from the water trough or other frequently used places.

Control Measures—Pesticides are poisonous and should always be used with caution. An excellent publication is "Control of External Parasites of Livestock," Leaflet 2854, by the Agricultural Extension Service, University of California. This publication summarizes the control of parasites on beef cattle as shown in the accompanying tables.

PARASITE CONTROL ON BEEF CATTLE AND DRY DAIRY STOCK

Chemical Formulation and Concentration			Application Method	Amount	Days Between Treatment and Slaughter (S) or Lactation (L)	Remarks
• HORN FLIES (Treat in warm months when flies are abundant.)						
Ciovap®	EC	10%*	Cattle backrubber	1 pt./2 gal. fuel oil (1%)	0	Ciovap®, coumaphos, methoxychlor, and ronnel may be applied as 1 gal. mixtures to 12 to 20 linear feet of burlap sacking wrapped around cable or chain. Recharge in 3 to 4 weeks. These chemicals may also be used in tank-type backrubber fitted with or without canvas drape or stringy mop.
coumaphos (Co-Ral®)	EC	11.6% (1 lb./gal.)		10 fl. oz./gal. fuel oil (1%)	0	
methoxychlor	EC	25% (2 lb./gal.)		1½ pt./gal. fuel oil (5%)	0	
ronnel (Korlan®)	EC	24% (2 lb./gal.)		5 fl. oz./gal. fuel oil (1%)	0	
coumaphos (Co-Ral®)	dust	1%	Dust bags	Add 2-3 lb. dust/bag. Recharge as necessary. Do not allow more than 1 oz./head-day use.	0	Coumaphos, malathion, and methoxychlor dusts may be used in dust bags made from narrow-gauge burlap sacks and grommeted at open end for hanging. They may be hung from a chain suspended between posts for free choice or mandatory daily use. Number of bags and placement depend on herd size, animal age, and habits.
malathion	dust	5%			0 (S) 14 (L)	
methoxychlor	dust	10%			0	
stirofos (Rabon®)	dust	3%		Use as directed.	0	

famphur (Warbex®)	ready mix	13%	Pour-ons by split applications for grub control.			See instructions, limitations of use, and remarks given for use of these animal systemics for cattle grub control.
fenthion (Tiguvon®)	ready mix	3%				
Ruelene®	ready mix	13.4%				
trichlorfon (Neguvon®)	ready mix	8%				

• FACE FLIES (Treat during late spring and early summer before flies become abundant.)

dichlorvos (Vapona®)	EC	21.8% (2 lb./gal.)	Brush or smear on face.	1 oz./gal. mineral oil or light fuel oil (0.25%)	0	One to 2 oz./animal head. Apply daily if needed.
coumaphos (Co-Ral®)	dust	1%	Dust bags	Add 2-3 lb. dust/bag. Recharge as necessary. Do not allow more than 1 oz./head-day use.	0	Dust bags made from narrow-gauge burlap sacks and grommeted at open end for hanging may be hung from a chain or board in front of feed or water troughs for mandatory daily use. Number of bags and placement depend on herd size, animal age, and habits.
malathion	dust	5%			0 (S) 14 (L)	
stirofos (Rabon®)	dust	3%			0	

Parasite Control on Beef Cattle and Dry Dairy Stock—continued

Chemical Formulation and Concentration			Application Method	Amount	Days Between Treatment and Slaughter (S) or Lactation (L)	Remarks
• CATTLE GRUBS (Heel flies or warbles) (Treatment during July through September preferred.)						
coumaphos (Co-Ral®)	WP	25%	Dip	8 lb./100 gal. water (0.25%)	0 (S)	See restrictions for coumaphos given for horn flies. Agitate dip mixture and adjust concentration before re-use.
	ready mix	4%	Pour on ½ oz./cwt. (no restriction on maximum dose).	Use directly from container.	0 (S) 14 (L)	See restrictions for coumaphos given for horn flies.
famphur (Warbex®)	ready mix	13%	Pour on ½ oz./cwt. but not over 4 oz./ animal. Split application possible.	Use directly from container.	35 (S) 21 (L)	Do not apply in conjunction with oral drenches or other internal medications or with other organic phosphates. Do not treat calves under 3 months of age nor sick animals or animals under stress from disease, overexertion, or excitement or within 10 days after shipping or exposure to disease. Cattle should have free access to water and feed before and after treatment. Follow all warnings and caution statements on the label. Famphur applied in winter months also gives good louse control.
fenthion (Tiguvon®)	ready mix	3%	Pour on ½ oz./cwt. (no restrictions on maximum dose). Split application possible.	Use directly from container.	35 (S) 28 (L)	

® Registered trade name.

* Contains 0.25% dichlorvos (Vapona®).

The information found in this table is representative of that contained in the publication by the University of California, Division of Agricultural Sciences, Leaflet 2854, entitled, "Control of External Parasites of Livestock." Other classes of livestock and a wide variety of control methods are included in this publication.

ABBREVIATIONS USED

gm.	gram(s)	tbsp.	tablespoon(s)
cc.	cubic centimeter(s)	/	per
fl. oz.	fluid ounce(s)	EC	emulsifiable concentrate
pt.	pint(s)	WP	wettable powder
qt.	quart(s)	sol.	solution
gal.	gallon(s)	psi	pounds per square inch
lb.	pound(s)	cwt.	100-lb. body weight
tsp.	teaspoon(s)		

GUIDE FOR MIXING SPRAYS

Insecticide, formulation,[1] and strength	Percentage of insecticide desired in spray	Amount of formulation to mix with water	
		100 gallons	5 gallons
ROTENONE:			
Derris or cube powder containing 5 percent of rotenone.	0.006	1 pound	5 tablespoons.
PYRETHRUM: EC, 1 percent [2]	.025	2½ gallons	16 ounces.
METHOXYCHLOR:			
EC, 25 percent	.5	2 gallons	13 ounces.
WP, 50 percent	.5	8 pounds	6½ ounces.
LINDANE:			
EC, 20 percent	.03	20 ounces	2 tablespoons.
	.05	1 quart	3 tablespoons.
WP, 25 percent	.03	1 pound	3 tablespoons.
	.05	1⅔ pounds	5 tablespoons.
TOXAPHENE:			
EC, 60 percent	.5	5½ pints	4½ ounces.
WP, 40 percent	.5	10 pounds	8 ounces.
MALATHION:			
EC, 57 percent	.5	1 gallon	6 ounces.
WP, 25 percent	.5	16 pounds	13 ounces.
COUMAPHOS:			
WP, 25 percent	.125	4 pounds	3¼ ounces.
RONNEL:			
EC, 24 percent	.125	2 quarts	3¼ ounces.
WP, 25 percent	.125	4 pounds	3¼ ounces.
CARBARYL: WP, 50 percent	.5	8 pounds	6½ ounces.
RUELENE: EC, 25 percent	.5	2 gallons	13 ounces.
CIODRIN: EC, 47 percent	.25	2 quarts	3⅓ ounces.
	.5	8⅓ pints	7 ounces.
DIOXATHION:			
EC, 47 percent	.15	2⅖ pints	2 ounces.
EC, 30 percent	.15	2 quarts	2½ ounces.

[1] EC=emulsifiable concentrate; WP=wettable powder.

[2] Refers to pyrethrins content. Pyrethrum formulations usually contain a synergist to improve effectiveness.

From Leaflet No. 456, U.S.D.A.

6. Controlling Internal Parasites

Different Types—Several different types of stomach and intestinal worms affect cattle, but the common stomach worm offers the greatest menace. Liver fluke is also a very serious internal parasite, especially in

Courtesy, Division of Parasitology, University of California, Davis

Fig. 7.10—Steer showing typical symptoms of having stomach worms. Notice swelling (bottle jaw) under throat and water discharge on tail.

mild climates. Animals most seriously affected are calves and yearlings; losses rarely occur among older cattle.

Symptoms—Symptoms commonly produced in calves are excessive scouring, loss of weight, general weakness, and anemia. The skin and mucous membranes are pale, the hair rough, and the animals often become "pot bellied." In later stages swelling may appear under the jaw, the so-called "bottle jaw" or "poverty jaw."

Preventive Measures—Good feeding conditions so animals are not licking the ground continuously help because worm eggs and larvae are not picked up so much.

Preventive measures for internal parasites include proper disposal of offal and carcasses, pure source of drinking water, and precautions taken so cattle do not have access to human feces. Calves should be kept off pastures known to be infected.

Keep Worm Population Low

The chance of contamination will be reduced if the parasites can be held in check. Rotating pastures and stocking at the proper rate will help. Irrigated pastures should be well drained. The use of temporary pastures where the seedbed is worked up cuts down the number of worms. Hay and grain should never be fed on the ground where worm problems are severe.

Control Measures—Much progress has been made in

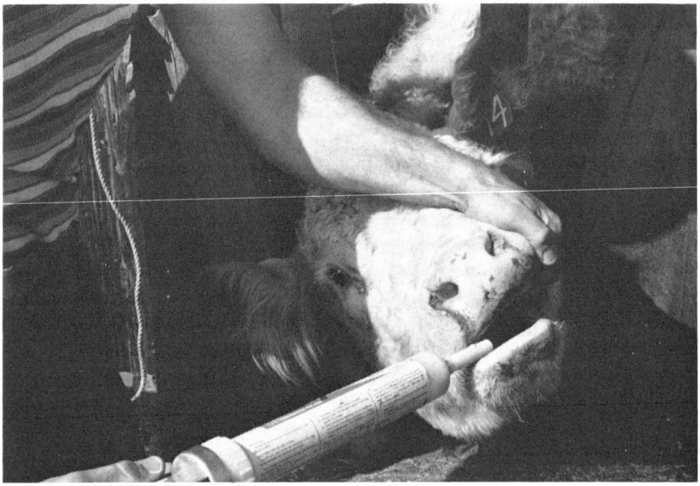

Fig. 7.11—Internal parasites can be controlled by periodic worming. This scene demonstrates the method by which worming paste is given to an animal. The compound is very sticky so the animal cannot spit it out but must swallow the paste. The worming gun illustrated holds approximately 20 doses. It is a fast, easy, uncomplicated way of individually dosing cattle.

recent years in controlling internal parasites. Standard remedies like copper sulfate, oil of chenopodium arsenicals, and so on still have their place, but newer chemicals in most cases are much more effective and control a wider range of parasites. Phenothiazine was one of the first breakthroughs and is still a desirable compound to use. Thiabendazole, Trichlorfon, Tramisol and so on are new chemicals known either by their chemical formula or by the trade name of the particular company producing the product. These compounds are slightly more expensive than older type wormers but much more effective and are an excellent investment. They are available at all supply houses and should be used only when directions recommended by the manufacturer are rigorously followed. The local extension agent, vocational agriculture teacher, and veterinarian can assist in local areas.

7. Controlling Anaplasmosis

Importance—Each year, anaplasmosis costs American cattle raisers millions of dollars. In some outbreaks it kills half the cattle that catch it; other times, less than 1 per cent will die.

All cattle can catch anaplasmosis; it is most severe in mature cattle and usually is mild in calves.

Anaplasmosis may be found almost anywhere that cattle are raised. It occurs most frequently in warmer areas and is spreading.

Cause—Anaplasmosis is a disease caused by microscopic parasites that live in red blood cells.

Cattle are more likely to catch anaplasmosis in the warm months of summer and early fall than at any other time. They can catch it at any time, however.

Anaplasma parasites attack red blood cells; this causes anemia. The anemia makes lips, nostrils, mouth

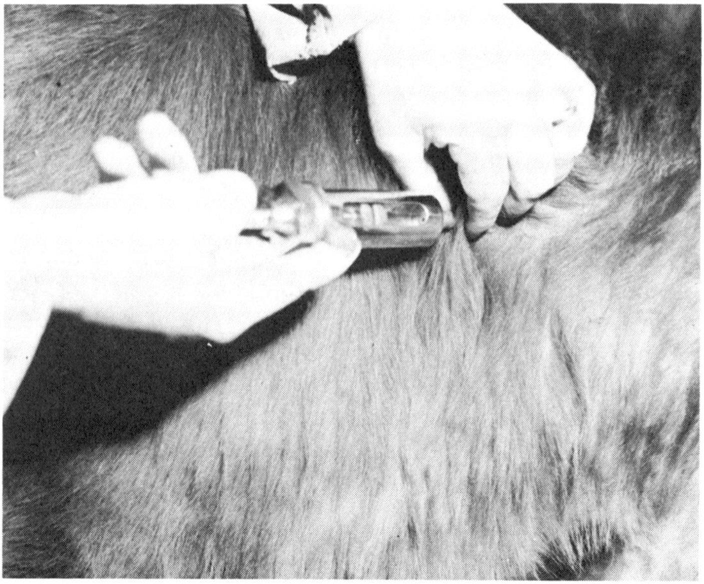

Fig. 7.12—Antibiotics and vaccines are often injected to prevent disease. The illustration shows how subcutaneous materials are injected. Livestock persons should learn this skill.

linings, and other visible mucous membranes look pale. It is spread when dehorning tools, vaccinating needles and so on are used on a carrier animal and then without sterilizing used on healthy cattle. It does not spread by contact or through feces, but can be spread by insects or ticks biting infected animals and then biting healthy stock.

Symptoms—The animal breathes faster than usual, and its heart beats faster. It loses condition. The animal grows weaker, stops chewing its cud, and is constipated. Symptoms vary in intensity; recovery is slow. When the disease lasts a long time, the animal's tissues become yellowish.

The active disease may last two to four weeks after

the first symptoms appear. It may disappear in one or two days. In rare cases, older cattle die within 24 hours. Sometimes the disease lasts a long time.

Animals that recover from the active disease are carriers and can spread the disease.

Control Measures—U.S.D.A. Leaflet 437 suggests the following practices:

a. Call a veterinarian as soon as you notice symptoms of anaplasmosis. He may give blood transfusions. These help the animal recover.

b. Use antibiotics, if indicated. Certain ones may shorten the course of the disease if used early.

c. Keep sick animals away from well ones. Keep them in the shade, and give them plenty of fresh water and a succulent feed.

d. Apply insecticides to both sick and well animals to protect them from biting insects and ticks. A veterinarian can blood test the herd so that carriers can be eliminated.

8. Controlling Anthrax

Importance—Anthrax, which is characterized by sudden death, causes losses of many millions of dollars annually to the livestock industry. As it is also infectious to man, anthrax is a menace to public health. Cattle, horses, and sheep are most commonly affected. The disease occurs at more or less definite times of the year, but generally in the summer. This period is referred to as the "anthrax season."

Cause—A germ known as the anthrax bacillus is the specific cause of the disease. The germ exists in the soil in certain regions which are known as "anthrax districts." These districts are widespread. The germ enters the body mainly with feed and water.

Symptoms—There are several forms of the disease, one of which is so acute, death may come in a matter of hours or even minutes with practically no other symptoms. In other forms, the animal lags, hangs head low, refuses to eat, and prefers to lie down. Respiration is rapid and labored. Body temperature is high and there are bloody discharges from the natural body openings.

Control Measures—U.S.D.A. Farmer's Bulletin 1736 suggests the following control measures:

First—Anthrax is a menace to the livestock industry and a qualified veterinarian or state official must be consulted.
Second—Prompt and proper disposal of all carcasses, especially those dead from the disease. Caution must be used in handling carcasses.
Third—Vaccination before the "anthrax season" in "anthrax districts" by a qualified veterinarian. Vaccination for anthrax is **not** 100% effective.
Fourth—Fence areas wherever practicable to exclude livestock from pools, swampland, buzzard roosts, and pastures where anthrax is known to occur.

9. Controlling Bang's Disease

Importance—It has been said that there is no disease of cattle in the United States more important from an economic standpoint than brucellosis or Bang's disease. Accumulated information shows that infected herds are as much as 30 per cent less profitable than clean herds. This figure is based on losses of calves, subnormal gains in beef calves, decreased milk production, and decreased sale of infected animals. It is doubly important because it is the source of undulant fever in man.

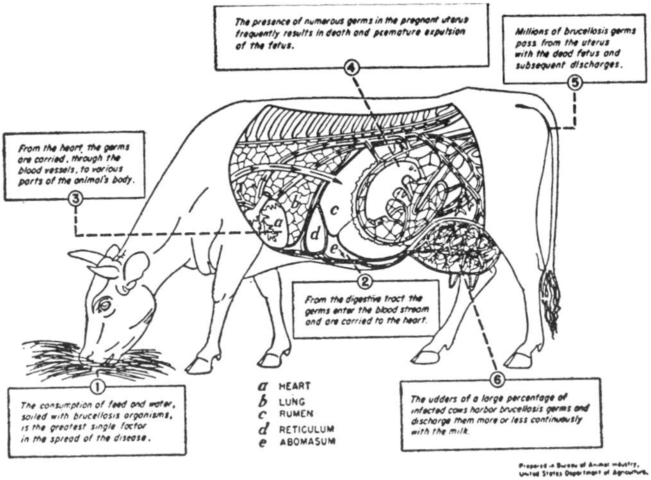

Fig. 7.13—Route of brucellosis germs in their attack on cattle.

Cause—The disease is caused by bacterium Brucella abortus, although there are other closely related forms. Cows and heifers of breeding age are most commonly infected. Calves are very resistant until breeding age.

Symptoms—The act of aborting is the only easily recognized symptom and this is not proof as other factors may cause abortion. Cows generally abort once and then seldom abort again. The common way of infection is through the mouth.

Control Measures—Drugs are of no value at present; therefore beef producers should:

a. Take a blood sample from each cow in the herd and have it tested for reactors. (A qualified veterinarian can best do this job.)
b. Remove reactors from the herd and sell for slaughter.

Courtesy, Cutter Laboratories, Berkeley, Calif.

Fig. 7.14—Seven-month fetus aborted from a cow infected with Bang's Disease. Aborting is the most easily recognized symptom.

c. Practice sanitation and prevent clean animals from coming in contact with infected animals.
d. Vaccinate calves with the proper strain of vaccine between the ages of 3 and 8 months.

Bang's disease, while important, is not so great a threat as once was the case, because preventative measures have been extant for several years.

10. Controlling Blackleg

Importance—Blackleg is one of the most infectious

diseases and one that usually proves fatal. It affects young cattle particularly.

Cause—A microorganism (clostridium chauvoei) is the causative agent. It can live only in the absence of oxygen.

Symptoms—The symptoms of blackleg are characteristic so recognition is easy. According to U.S.D.A. Yearbook, 1956, at first there is high fever, loss of appetite, suspension of rumination, and great depression. Breathing becomes rapid and the animal lies down. Rapidly developing swellings or tumors under the skin are characteristic. Upon pressure, the tumors

Fig. 7.15—This cow had blackleg; while the rope appears to have been restraining her, such was not the case. She was weak in the rear quarters and was continually on the brink of falling. Even though attempts at treatment were made, she died. Prevention via vaccination is the only way of controlling the disease.

make a peculiar crackling sound due to gas formation by the germ. Death generally occurs in 12 to 36 hours after the first appearance of the disease.

Control Measures—The only reliable and effective control measure against blackleg is vaccination. Once the symptoms appear, vaccination will be of no value. Many products are available and give excellent results. It must be remembered that a vaccine is a preventative and not a curative agent. Therefore, calves should be vaccinated before 6 months of age. Should an outbreak occur healthy animals should be vaccinated or revaccinated.

Courtesy, Cutter Laboratories, Berkeley, Calif.

Fig. 7.16—This calf has just died from blackleg. Note how animal has begun to stiffen up immediately due to production of gas under skin.

11. Controlling Choke

Importance—Choke, itself, is not particularly dangerous; however it may lead to bloat with eventual loss of the animal.

Cause—Choke in cattle is caused by an attempt to swallow large objects without thoroughly chewing them. Such food materials as beet tops, ears of corn, various root crops, and potatoes are the principal causes of choke and it is occasionally caused by foreign objects.

Symptoms—The symptoms in most cases are quite alarming, but sudden death seldom occurs. However, restlessness, salivation, retching, forced swallowing movements, and eventual bloating are indicative of choke.

Control Measures—The object may sometimes dislodge in one or two days without mechanical assistance of any kind. According to U.S.D.A. Yearbook, 1942, rough procedures like trying to crush the object between two bricks should not be used. Rather, if the object is an article of food lodged in the neck portion of the esophagus, it should be forced upward as gently as possible to the pharynx at the back of the mouth cavity by placing a hand on either side of the neck along the jugular furrow and pressing upwards against the object. If the object is lower down in the throat, it should be forced on down into the rumen. This can best be done with a flexible stomach tube about 1½ inches in diameter which has a wooden probang inserted in the end. It should be first lubricated with linseed oil. If neither of these procedures gives results, a veterinarian should be called.

12. Controlling Foot Rot

Importance—Foot rot, or foul foot, is an all too common disease around muddy corrals or infected lots. It is seldom found in cattle on the range, but is often a real problem where cattle are concentrated in a feed lot or grazing on irrigated pasture.

Cause—The disease is apparently caused by one or more microorganisms, although authorities have not at present agreed upon the exact germ or germs responsible. It seems to be anaeorobic, meaning it can't survive in the air.

Symptoms—Foot rot normally begins with an injury to the skin and then infection. Cattle become sore footed and the infected foot becomes swollen and foul

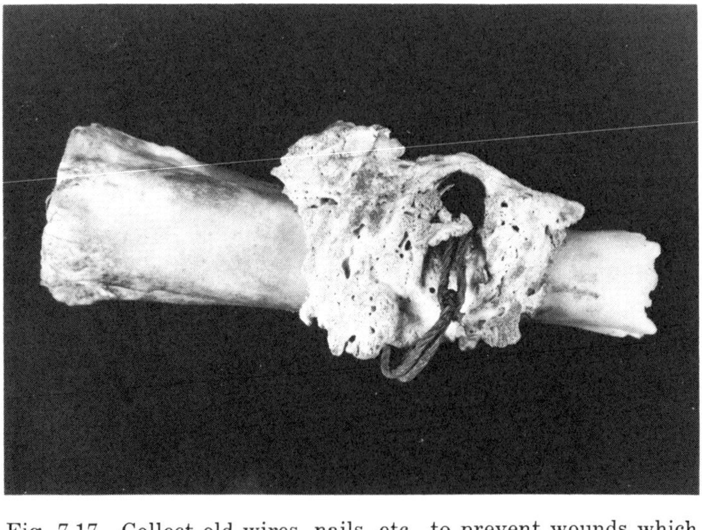

Fig. 7.17—Collect old wires, nails, etc., to prevent wounds which allow germs to attack. This wire somehow completely encircled the leg of the calf and caused a severe wound, with ultimate crippling, even though the bone grew a protective shell and the animal lived.

Courtesy, Jensen-Salsbery Laboratories

Fig. 7.18—Respiratory diseases are becoming increasingly important. The plastic canula used for administration of the new *TELC*™ strain intranasal Parainfluenza-3 vaccine, Nasalgen-P, is shown being inserted in this vaccination scene. When the canula is fully inserted, vaccine is sprayed on the walls of the nostrils. A standard vaccinating syringe is used.

smelling. The disease occasionally is under the horny part of the foot and not readily visible. Lameness is an obvious symptom.

Control Measures—According to Iowa Extension Bulletin, foot rot can be cleared up by proper treatment:

a. Keep corrals and feed lots as dry as possible.
b. Clean up all boards with nails, wire, etc., that could injure the feet of the animals.
c. Keep feet of animals properly trimmed.
d. Fix a shallow vat or tank in the yard so the cattle must stand in it when they drink or go to feed. The tank should contain a saturated solution of copper sulfate. A second procedure is to prepare a 1 to 20 mixture of powdered copper sulfate and slacked lime which is spread around water troughs or where mud and manure accumulate.
e. In severe cases consult a veterinarian, as sulfa injections frequently give good results.

13. Controlling IBR

Infectious bovine rhinotracheitis (red nose) is a disease of the upper respiratory passage. Cattle are most susceptible to the disease under confined conditions. It was first encountered in Colorado feedlots in the 1950's. Death loss may run from 3 to 5 per cent, but serious loss of weight is the most significant problem.

Cause—It is caused by a specific virus and shows up generally after animals are on feed and taking on flesh.

Symptoms—Affected animals exhibit excessive nasal discharge, high temperature, rapid and difficult breathing, coughing, and depression, and go off feed. Pregnant animals can abort. The disease could easily be confused with shipping fever or other respiratory diseases, so careful diagnosis is required.

Control—Affected animals should be isolated immediately upon recognition of symptoms and treated with antibiotics or sulfas to subdue secondary infections. Routine vaccination is the only preventative means. Do not vaccinate pregnant animals. A veterinarian should be consulted.

14. Controlling Leptospirosis

Importance—Leptospirosis, commonly called "lepto," is a widespread disease of all domestic animals, especially cattle and swine. It ranks as one of the most serious threats from an economic standpoint, so should be understood. While death loss is low, abortion and loss of weight are high.

Cause—The disease is caused by a spiral, thread-like bacterium. In the United States the most common strain is pomona. Animals get the disease by taking the organism into their bodies in the food or water they eat, in the air they breathe, or from open wounds. It can spread from one kind of animal to another (e.g., cattle and deer).

Symptoms—Mild "lepto" may escape notice completely although pregnant animals may abort and, worst of all, may shed the organism and spread the disease. Severe cases show sudden fever (103° to 107°F.), depression, lack of appetite, rapid breathing, and pale, yellowish color on eyes and mouth. Milk will be bloody. Pregnant animals abort and shed afterbirth with the calf. Recovery is slow.

Control—The only effective means is through vaccination. However, if "lepto" is suspected and cattlemen wish to maintain a herd free of the disease without vaccination, blood tests should be used, especially on new animals. Blood transfusions may help. Antibiotics and good care help. Consult a veterinarian.

15. Controlling Pinkeye (Infectious Keratitis)

Importance—The disease is not fatal unless from secondary causes; however, animals go down hill and become thin. An entire herd may become infected unless adequate precautionary measures are taken. Temporary blindness in one or both eyes results and occasionally an animal may go completely blind. Calves especially are susceptible. Some breeds may be more susceptible than others.

Cause—Even though the disease has been known for a long time, the exact cause is not known except that it is caused by microorganisms. It is extremely contagious and spreads from mucus secretions from the eye of one animal to other susceptible animals. Dust and insects, especially in warm weather, favor the rapid spread of the disease.

Symptoms—Eyelids swell and eyes have a watery discharge at first, which later contains mucus and pus. Lining of lid becomes red and congested. The animal displays the usual signs of uncomfortableness, listlessness, loss of appetite and weight.

Control Measures—Many remedies have been tried including some cruel ones like throwing sand or salt in the eye. No specific control is known but some veterinarians have had success with vaccination or an injection of antibiotics into the eyelid. A veterinarian should be called if the disease appears at all serious or widespread.

Wet Down Working Areas—Corrals, barn floors,

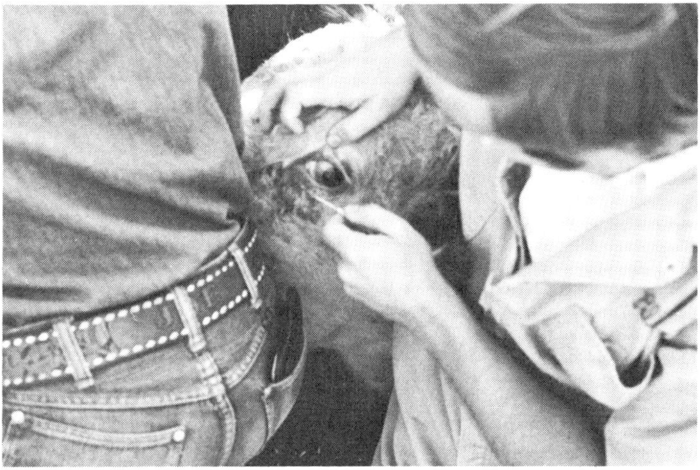

Fig. 7.19—Treating cow for pinkeye. Notice that the operator is injecting the compound into the eyelid of the affected eye in order to effect more rapid recovery.

transportation vehicles and any place where cattle and especially calves are to be worked should be sprinkled down with water prior to letting the animals enter. The germs seem to be spread by wind and dust so this is especially important during the summer months or dry dusty periods. Generally speaking, the disease is no problem during the winter.

A desirable practice is to attempt the following at the first sign of an outbreak:

a. Separate infected animals and put them on grassy pasture if possible.
b. Keep them out of the wind and dust.
c. Control flies and insects, especially on the infected animals.
d. Use spraybomb containing an antibiotic, spraying directly into eye and surrounding eyelid.
e. Provide easy access to feed and water as

animals cannot search out their own. Do not harass them.

f. Call a veterinarian for possible vaccination if the situation seems critical.

16. Controlling Red Water Disease

The death rate is high (exceeding 90 per cent) ; however, at present the disease is not widespread. Urine from sick animals gives the disease its name, red water, because it is dark red or port wine in color. However, it is easily confused with other anemia-producing diseases, such at anaplasmosis and leptospirosis. It might also be confused with anthrax, blackleg, or hemorrhagic septicemia, because it causes quick death.

Cause—The causative organism is a spore-bearing bacterium which enters the body of the animal through food and water. Once in the animal's body it liberates a powerful toxin which destroys the red blood cells. The disease is acute and infectious.

Symptoms—Appetite, rumination, bowel movements, and lactation suddenly cease, and breathing becomes shallow and difficult. High temperature, up to 106°F., is noticed early; however, it soon recedes, becoming subnormal just before death.

Control—Early seasonal vaccination will prevent the disease. However, a veterinarian should be consulted to see whether or not the disease is prevalent and vaccination is recommended. On badly infested land, animals may have to be revaccinated several times.

17. Controlling Scours

Importance—According to Oregon State College Circular 154, diseases of calves accompanied by scours are responsible for the greatest losses among these animals. Losses of calves on the range are in most cases equivalent to losses of mature cattle, as feed is available and labor required to raise the remaining calves is little less than that required to raise a full crop.

Cause—A variety of factors may cause scours. Too rich milk, sudden changes in feed or feeding habits may cause mild scouring. Parasites, such as worms, may cause a serious type of scours known as Black Scours. White Scours, or infectious diarrhea, is a very serious and often fatal type of scours that is caused by a germ "colon bacillus." Other microorganisms may be responsible for bloody scours.

Symptoms—The most constant symptom is a severe diarrhea. Hind parts appear dirty and calves strain. There is loss of flesh and general weakness. Symptoms vary, and if more than a mild case develops, a veterinarian should be consulted. A common symptom of white diarrhea is a dirty white colored, fluid-like, frothy discharge with an extremely offensive odor.

A common symptom of bloody scours is blood visible in the feces produced by severe diarrhea.

A common symptom of black scours is a more or less dark-colored diarrhea.

Control Measures—Proper sanitation is the key to prevention of scours. Calves should not be fed on the ground, and if possible should not be allowed to eat food which is contaminated with droppings. Keep calves on clean pastures. Range calves are seldom infected as often as calves raised in and around a barn. Proper nutrition is important, as well-fed calves generally are more resistant. In cases of black scours, drugs such as tetrachlorethylene or phenolthiazine should be administered to eliminate worms. In cases of white scours, the feeding of acidophilus milk has given good results. Some relief from diarrhea may be had by using the same compounds humans use for diarrhea.

Treatment—After all preventive measures have failed, treatment is a serious problem even for a veterinarian. The U.S.D.A. Yearbook says that much can be accomplished by blood transfusions and modern antibiotics and chemicals. The Pennsylvania State University reports that the administration of Aureomycin was very helpful in preventing scours and in reducing losses in those animals which have scours.

A good vaccine may be available in the future which would satisfactorily prevent the disease.

18. Controlling Shipping Fever

Importance—Shipping fever, or stockyard fever, is an infectious disease of cattle usually attended with a high mortality. Its medical name, "hemorrhagic septicemia," was given because it is a poisoning of the blood. The disease occurs in all parts of the United States, especially in under-nourished cattle and young cattle. Outbreaks are generally associated with shipment of cattle from one point to another or any other form of stress.

Cause of the Disease—The primary cause has not been definitely established. However, the germ Pasteurella boviseptica is present in animals with the disease. Many investigators believe there are two factors (virus and bacterium) present as necessary causitive agents. The disease is known to be highly infectious.

Symptoms—The disease develops rapidly and lasts from 2 to 8 days or longer. High temperatures, 104-107 degrees F., are accompanied by loss of appetite, occasional coughing and watery discharges. Weakness and diarrhea follow. Animals may develop pneumonia and die in 48 to 72 hours. Some animals recover. In any case, symptoms generally develop after the animal has gone through some hardship, such as a long truck ride with little attention. Recovery is slow.

Control Measures—Many control measures could be enumerated, but all would revolve around eliminating stress and hardship for the animal; however, to mention some:

a. Avoid hard driving and allow ample time for rest before loading. Brand, castrate, etc., at least 30 days prior to shipping.
b. Avoid over-crowding in cars or trucks. Cool or insulate cars as climate demands.
c. Give feed and water at proper intervals en-route.
d. Do not withhold water so cattle will take on too large a fill. Supply a non-legume hay.
e. Upon arrival, withhold water at first, but supply ample dry roughage.
f. Do not make sudden changes in feed.
g. Segregate diseased animals.

The use of biological products is at present a controversial subject. Therefore, their administration should be left to a competent veterinarian. However, many stockmen feel the use of bacterins is justified if given several weeks prior to shipment.

Bacterins should not be given after shipment although antibiotics may be helpful at this time.

19. Controlling Tetanus

Tetanus is a serious disease, as death occurs in almost 85 per cent of the cases. All animals and all ages are susceptible. It is usually associated with a wound, as the disease is caused by an anaerobic bacterium. The germ releases a powerful toxin (more than 100 times as toxic as strychnine). Therefore, dehorning, castration, and so on, provide possible modes of entry, although the most perilous situation exists from deep unnoticed wounds.

Symptoms—The first sign of tetanus is stiffness about the head. The animal shows difficulty in chewing. The third or the inner eyelid may protrude over the forward surface of the eyeball. With the slightest movement or noise, the animal may show a violent spasm. Affected animals usually remain standing until close to death.

Prevention—Vaccination with tetanus toxoid is the safest and best procedure. The toxoid produces an ac-

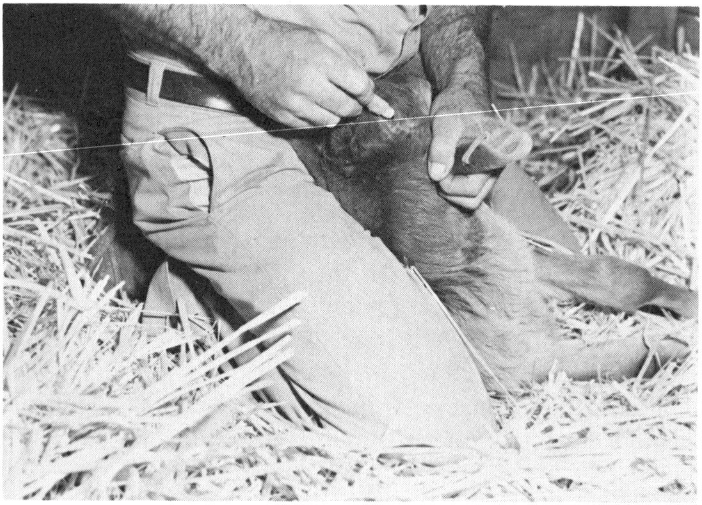

Fig. 7.20—One good way to remove horn buttons is by use of caustic or hornpaste. Horns removed early seldom cause problems later.

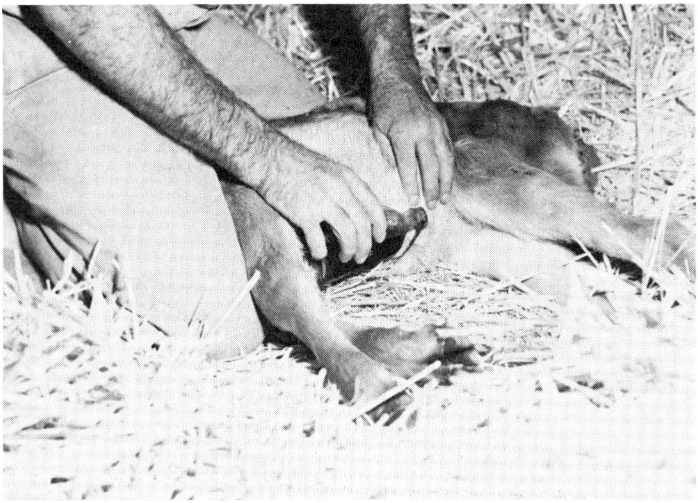

Fig. 7.21—Proper way of applying iodine to calf's navel. This practice is part of the program to prevent tetanus and other diseases.

tive immunity by stimulating a response in the injected animal which causes it to build its own long-term immunity. At present, few cattlemen routinely vaccinate for tetanus. This procedure may be changing, as many alert stockmen now consider it an approved practice and vaccinate whenever they observe a deep wound, a lame foot that may result from stepping on a nail, or any other suspect of a condition where anaerobic bacteria could live.

Treatment—Once it is apparent an animal has tetanus, treatment is generally useless and uneconomical. However, tetanus antitoxin may be helpful if administered at the time of the first symptoms. Animals should be kept quiet, and a veterinarian should be consulted.

20. Controlling Trichomoniasis

This is a venereal disease transmitted by bulls. As yet it is not widespread, but many livestock men believe it is becoming more prevalent.

Cause—A parasite (Trichomonas fetus) is the causative organism found particularly in the sheath of bulls. It is transmitted either by natural or artificial breeding.

Symptoms—Abortion early in pregnancy. Females are difficult to settle with calf, and have irregular heat periods. Diagnosis is difficult so a veterinarian should be called if the disease is suspected.

Control Measures—A rest period of two or three heat periods may help females overcome the infection. Treatment of bulls is only partly effective. Artificial insemination from a clean bull may be used to prevent transmission of infection.

Unless animals are very valuable they should be culled and an attempt made to keep the herd clean.

21. Controlling Vibriosis

Importance—Like trichomoniasis this is a disease infecting the reproductive organs. It may become locally important and should be suspected whenever breeding difficulties are encountered.

Cause—The disease is transmitted at the time of breeding as infected bulls carry the organism. Causative agent is a microorganism (Vibrio fetus).

Symptoms—Whenever repeated matings fail to cause pregnancy the disease should be suspected. In the pregnant cow abortion and retained placentas result. Generally less than 50 per cent of the herd conceive at first matings if the disease is present.

Control Measures—Most individual cows will recover. Bulls are difficult to cure and should be disposed of unless very valuable. A veterinarian should be consulted.

22. Controlling Warts

Importance—Warts are a common occurrence in cattle. Occasionally, particularly on young cattle, they may become large and pendulous and sap the strength enough to stunt the growth of the animals. Their chief damage, however, is to calf skins and cattle hides after tanning. Cattle buyers make discounts for warty animals purchased in the markets. Purebred cattle have their appearance materially reduced if they are warty.

Cause—Common warts are a specific type of skin over-growth, non-malignant in character. The causative agent is a filterable virus and can be infectious.

Symptoms—Warts are generally found in younger cattle. Cauliflower-like growths appear about the head and neck which vary greatly in size and shape. They may be hard or soft, and occasionally show a tendency to bleed.

Control Measures—According to the U.S.D.A. Yearbook, all warty cattle, particularly calves and yearlings, should be removed from the herd. Rubbing posts, pens, chutes, etc., should be disinfected. Warts occasionally disappear without treatment, especially as animals get older, but generally treatment is necessary. Warts that are small at place of attachment may be removed by clipping with a sterile pair of scissors or tying a slender cord tightly about the base of the wart. The stumps of warts should be touched with tincture of iodine. Before warts become excessively large, they may be destroyed by daily applications of glacial acetic acid or tincture of iodine.

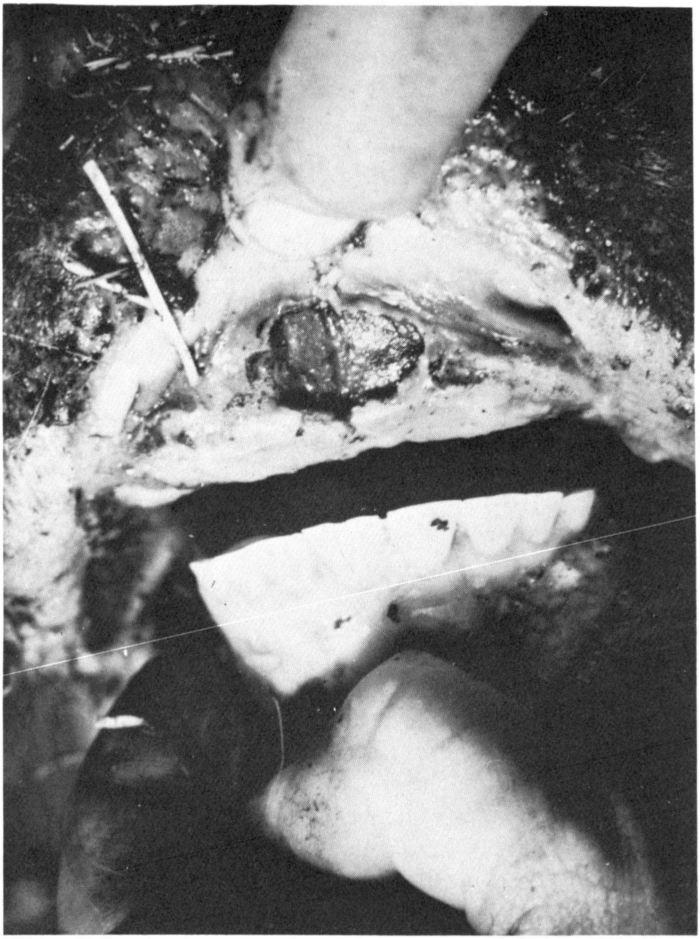

Fig. 7.22—Mouth of a cow affected with hoof and mouth disease. Notice the large lesion on the upper gums. Animals refuse to eat, become weak, and die from a variety of causes.

There is an internal remedy and a vaccine which are effective in wart control. Veterinarians should be consulted for this treatment.

23. Controlling Other Serious Diseases

Other diseases of beef cattle that can be serious are tuberculosis, foot and mouth disease, lumpy jaw, mastitis, cancer eye, pneumonia, and other less well-known maladies.

Importance—Actually the above mentioned diseases are some of the most serious diseases known to affect

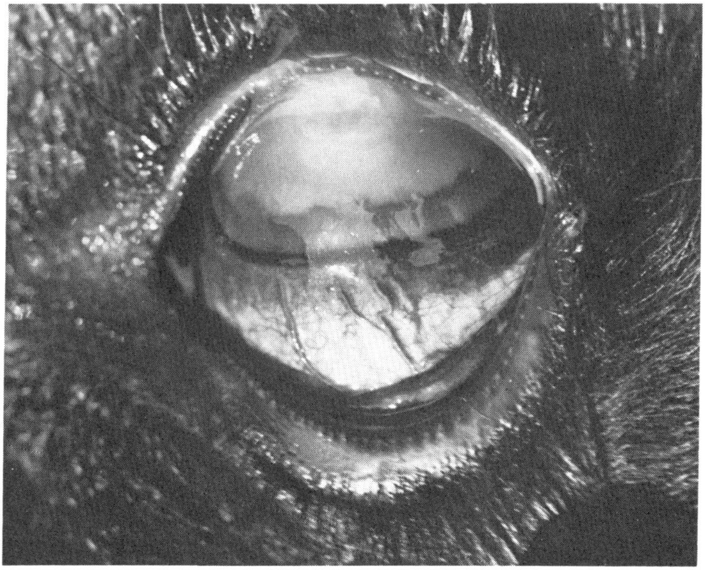

Courtesy, Cutter Laboratories, Berkeley, Calif.

Fig. 7.23—Clinical picture of pinkeye. Animals so infected should be separated from the rest of the herd immediately, and a veterinarian should be consulted. Traditional cures such as throwing sand in the eyes should not be used. Sulfa compounds have been highly effective.

cattle. Because of their importance, the government and livestock industry have taken cooperative steps to stamp out many of these diseases so that at present they do not constitute a serious threat to the individual producer. Nevertheless, they constitute a tremendous potential danger against which cattle raisers should constantly guard. In some local areas they may be a current danger.

Prevention of Diseases—Following are some suggestions which may tend to lessen the danger from these serious diseases becoming established.

a. Support and cooperate with local and governmental agencies engaged in controlling and preventing diseases, such as tuberculosis.
b. Immediately report any unusual illness of cattle to a veterinarian or the proper authority.
c. Practice proper sanitation in all cattle operations.
d. Remove objects, such as barbed wire, old nails, broken gates, etc., on which animals can injure themselves and allow infectious germs to enter their bodies.
e. Provide proper equipment so that cattle can be easily handled without injuring themselves.
f. Keep livestock well fed and healthy so they can better resist parasites and germs.

24. Treating Wounds

Remove Cause—All protruding objects, sharp corners and especially old barbed wire or boards with nails should be collected and hauled away.

Clean Wounds—Prompt cleansing and application of antiseptics to accidental wounds is helpful. Keep stock uncrowded on dry clean pasture so they are not overactive and give nature a chance. Most authorities state that many of the antibiotics like penicillin and Aureomycin used as an ointment or injected by someone competent seem to help.

Vaccinate for Tetanus—This is a simple inexpensive practice. Many ranchers routinely vaccinate for tetanus whenever surgery such as castration or dehorning is involved. It is especially important with wounds. Deep wounds may need a veterinarian's care.

NORMAL TEMPERATURE, PULSE RATE, AND RESPIRATORY RATE OF DOMESTIC ANIMALS

Species	Normal Temperature °F.	Pulse Rate per Minute	Respiratory Rate per Minute
Cattle	101.4	40-70	10-30
Horses	100.2	32-44	8-16
Pigs	102.2	60-80	8-18
Sheep	103.2	70-80	12-20

Chapter

VIII

BUTCHERING BEEF ON THE FARM

Select Proper Type—Yearlings from good type beef animals are most satisfactory for farm slaughter. Yearlings produce light carcasses and light cuts which chill quicker and the meat is more tender than that obtained from mature animals.

Quality meat can only be obtained from animals which have been properly fattened. Fat improves the tenderness and flavor and increases the juiciness of the meat. Meat which is well covered with fat keeps better and does not dry out as quickly as meat which has little or no fat.

Assemble Proper Equipment—For proper slaughtering a means of hoisting the carcass must be provided, such as a tripod made of timbers, a tree or a brace extending out from a building. A chain hoist, block and tackle, or windlass is necessary for hoisting. A pair of ordinary clevises should be used to fasten the tendons to prevent the carcass from slipping.

Use Good Tools—A good set of tools will help you do a quick and efficient job of butchering. Tools should be kept sharp and clean. A skinning knife and steel

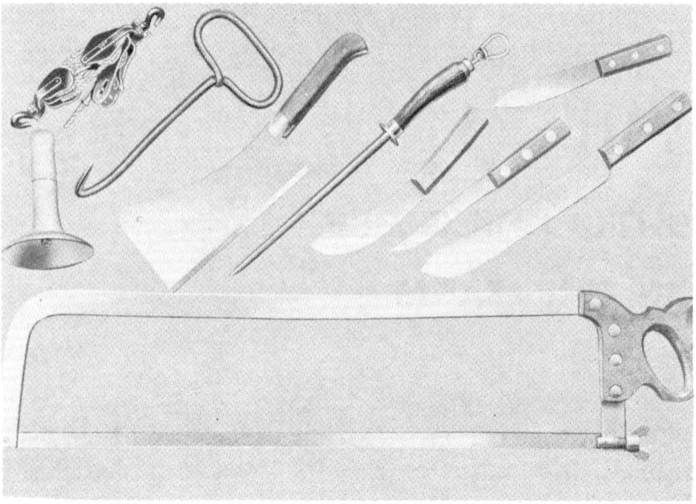

Courtesy, Morton Salt

Fig. 8.0—A good set of butchering tools should consist of sticking knife, skinning knife, boning knife, butcher knife, steel, cleaver, bell scrapers, meat saw, and meat hooks. Additional instruments that are very useful are a meat pump, thermometer, and meat needle, for sewing rolled cuts.

are essential. If a cleaver and meat saw are not owned, a hatchet and wood saw will serve. Plenty of buckets, tubs, cloths, and fresh water should be at hand.

Keep Animal off Feed—A better bleed will be secured and cleaning made easier if the animal is kept off feed for 24 hours before slaughtering, but it should be given plenty of water. Also, the animal should be kept quiet for some hours before killing and should never be killed when in an excited, overheated, or feverish condition.

Properly Kill Animal—Either shooting or stunning may be used for killing beef. The animal should be securely snubbed from a tree or post and stunned with a sharp blow from a sledge or axe. Draw imaginary

• CHUCK *164.8 lbs (26.8% of total carcass)*

	Saleable Beef—lbs	Other lbs
Blade pot roasts	59.3	
Stew or ground beef	32.1	
Arm pot roast	22.3	
Cross rib pot roast	10.7	
Boston cut	9.9	
Fat and bone		30.5
TOTAL	134.3 lbs	30.5 lbs

• BRISKET *23.4 lbs (3.8% of total carcass)*

Boneless	9.4	
Fat and bone		14.0
TOTAL	9.4 lbs	14.0 lbs

• SHANK *19.1 lbs (3.1% of total carcass)*

• RIB *59.0 lbs (9.6% of total carcass)*

	Saleable Beef—lbs	Other lbs
Standing rib roasts	24.2	
Rib steaks	12.4	
Short ribs	4.7	
Braising beef	2.7	
Ground beef	3.5	
Fat and bone		11.5
TOTAL	47.5 lbs	11.5 lbs

• LOIN *105.8 lbs (17.2% of total carcass)*

Porterhouse steak	18.7	
T-bone steak	9.5	
Club steak	5.2	
Sirloin steak	41.4	
Ground beef	2.9	
Fat and bone		28.1
TOTAL	77.7 lbs	28.1 lbs

Prepared by the National Live Stock and Meat Board

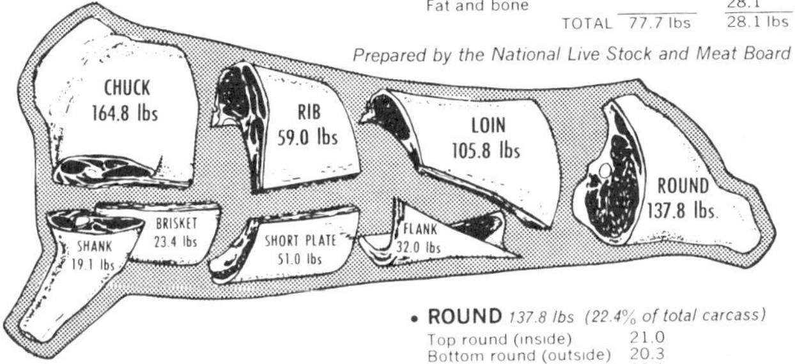

• SHORT PLATE *51.0 lbs (8.3% of total carcass)*

Plate, stew, short ribs	40.8	
Fat and bone		10.2
TOTAL	40.8 lbs	10.2 lbs

• FLANK *32.0 lbs (5.2% of total carcass)*

Flank	3.2	
Ground beef	12.6	
Fat		16.2
TOTAL	15.8 lbs	16.2 lbs

MISC. *22.1 lbs (3.6% of total carcass)*

Kidney, hanging tender	3.6	
Fat, suet, cutting losses		18.5
TOTAL	3.6 lbs	18.5 lbs

• ROUND *137.8 lbs (22.4% of total carcass)*

Top round (inside)	21.0	
Bottom round (outside)	20.3	
Tip	13.1	
Stew	8.3	
Rump	4.8	
Kabobs or cubes	2.1	
Ground beef	14.2	
Fat and bone		54.0
TOTAL	83.8 lbs	54.0 lbs

SUMMARY
(1000 lb choice steer)

Dresses out 61.5%	**615 lbs**
Less fat, bone and loss	**183 lbs**
Saleable beef	**432 lbs**

Fig. 8.1—The illustration indicates the origin and location of the various wholesale sections in a beef carcass as well as most retail cuts. Today's modern type 1,000-pound choice steer produces approximately a 615-pound carcass. The retailer then trims away about 183 pounds of fat, bone, and waste, ending up with only 432 pounds of beef he sells to customers.

cross lines from the horns to the eyes and stun at a point just above where these lines cross. This is the same location to aim for when shooting.

Stick on Ground—One of the sticker's legs should be placed in front of the animal's forelegs and the other placed back of the jaw. The sticker should stand with his back to the animal's neck to be protected from any sudden movements.

For sticking, a long incision is made along the center of the throat. The knife blade is pushed just under the point of the breast bone, severing the branching veins and arteries at that point.

Bleed in Pasture—In many rural situations it may be more practical to kill the animal, bleed it out, and remove the offal in a clean pasture far away from the farmstead. This eliminates a messy job in disposing of unwanted material, as nature quickly recycles it via predators and insects.

Pump Out Blood—After the animal is stuck, to hasten bleeding place one foot on the animal's flank and by alternately lunging forward on this foot and then pulling back on the tail the bleeding will be materially speeded up.

Skin Carefully—The hide is the most valuable byproduct of the beef and care should be taken to make a good "pattern" free from cuts and scores. A smooth job of skinning should be done to keep any meat from being left on the hide.

Skin the Head First—The hide is cut across the top of the head, then down to the eye and on to the corner of the mouth. In this way the skin may be removed from the face in one piece. Then start skinning at the throat where the sticking cut was made and carry the cut forward to the point of the jaw and the lower portion of the head. The head can be skinned out easier if it is turned up on its base.

Remove the Head—When skinned, the head is grasped by the lower jaw and unjointed at the first joint or "puzzle" bone. The head should be washed and the cheek meat and tongue removed, after which the skull can be split and the brain removed.

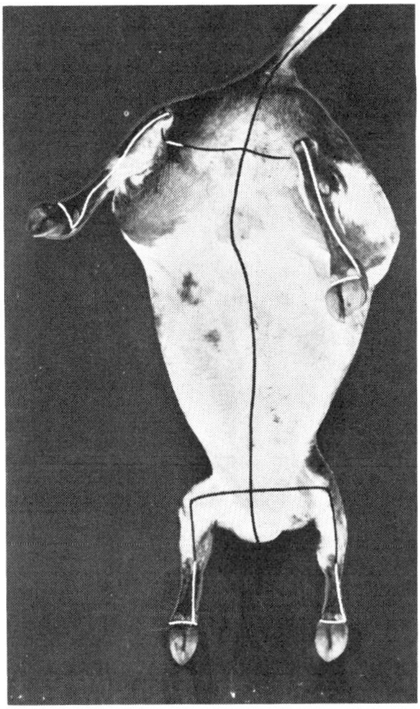

Courtesy, Morton Salt

Fig. 8.2—Different ripping cuts to begin taking off the hide.

Use Care in Siding—The most difficult part of the skinning operation is "siding." By pulling the hide upward and holding it tight and keeping the blade of the knife against the hide, cuts and scores can easily be avoided. The carcass should be sided down as far as possible since it is easier to do this here rather than delay the siding until the carcass is hoisted.

Loosen Gullet and Windpipe—The throat is split and gullet and windpipe worked out as far as possible. The gullet should be "tied off" with a stout cord.

Split the Aitch Bone—The belly wall is ripped by "choking down" the knife and using the fist to keep the viscera away from the cutting edge. This cut is

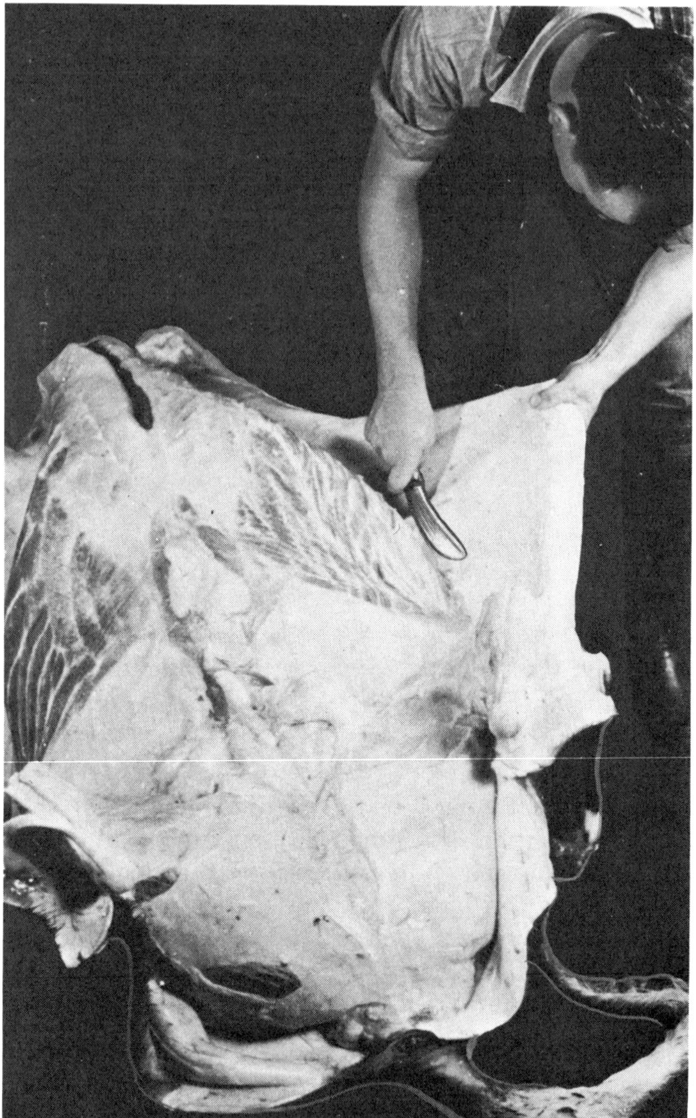

Courtesy, Morton Salt

Fig. 8.3—Removing the hide. Notice how the operator pulls upward with one hand while he keeps the blade of the knife flat against the hide.

carried backward to the aitch bone and the round split apart at this point. If the cut is accurately centered the pelvic or aitch bone can be split with a knife. With mature cattle a saw may be needed.

Remove Abdominal Viscera—First, loosen the rectum by cutting around it. After the rectum is cut free, it should be tied off with heavy twine and then worked loose from the backbone, leaving the kidney and bed fat in the carcass. Work the small intestines loose and let them drop down over the paunch. Have the liver washed thoroughly and placed in cold fresh water.

Cut Through Diaphragm to Remove Heart and Lungs—Between these two muscular portions is a wide connective tissue portion which must be trimmed out. The lungs and heart may now be removed. The edible offal should be trimmed, washed, and chilled. The liver is attached to the right side and the kidney hangs squarely under the backbone. It will be necessary to cut around it, leaving this organ on the left side along with the "hanging tender."

Complete Backing and Splitting—After the viscera are removed, hoist the carcass and finish removing the skin from the back. This is known as "backing." The carcass is now ready to be split down the center of the backbone. Although this is done commercially with a cleaver, most people can do a better and smoother job with a saw.

Wash and Shroud—After the carcass is split it should be washed with clear, tepid water and shrouded. A clean muslin cloth wrung out of hot water may be tightly pinned to each side of the carcass. This smooths out the fat and bleaches it. This is best accomplished at a temperature of 32 to 34 degrees.

Properly Ripening Beef—When meat is to be used

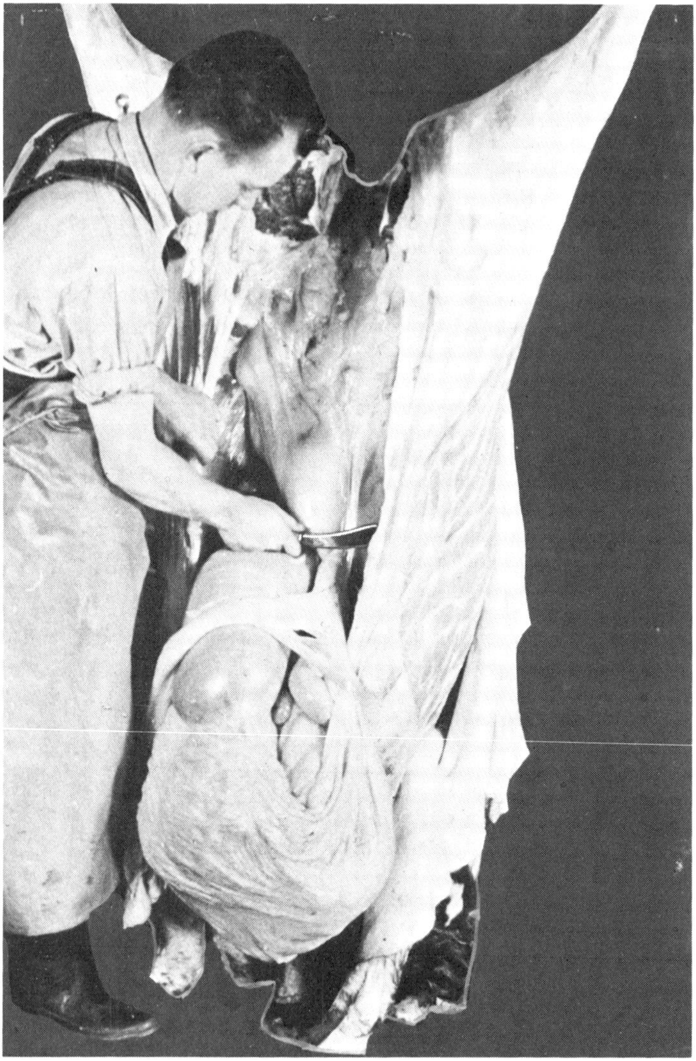

Courtesy, Morton Salt

Fig. 8.4—Removing the entrails. Care should be taken so as not to cut into the intestines.

fresh, the best meat is obtained if the beef is allowed to "ripen" from seven to ten days and, if possible, held at a temperature of 32 to 36 degrees during this time. This will improve the tenderness of the meat and also add to its flavor.

Cutting Up the Beef Carcass

Quarter the Carcass—The first step in cutting the beef carcass is quartering, which is dividing the fore and hindquarters. The carcass has thirteen ribs on each side. The first cut is made between the last two ribs, leaving twelve ribs on the forequarter and one rib on the hindquarters. Insert the knife between the twelfth and thirteenth ribs at the belly end of the ribs and make the cut all the way to the backbone. Then saw the backbone in two.

Cutting Up the Fore Quarter

Separate Plate from Rib—Measure 10 inches from the backbone or chine along the 12th rib. Using this as a starting point, a straight line cut is made, continuing the cut across the shank just above the elbow joint.

Saw Off Plate and Fore Shank—After the cut is made with a knife, the ribs are sawed through and the fore shank sawed off at the elbow.

Separate Rib and Chuck—Make a cut with the knife between the fifth and sixth ribs to separate the rib cut from the chuck. Complete the cut with the saw, sawing through the backbone.

Complete the Rib Cut—The rib cut is one of the choice forequarter cuts and comprises about 10% of the carcass. It is one of the more tender cuts and is extensively used for roasting. Roasts are made by cutting between the ribs. These roasts may be boned and

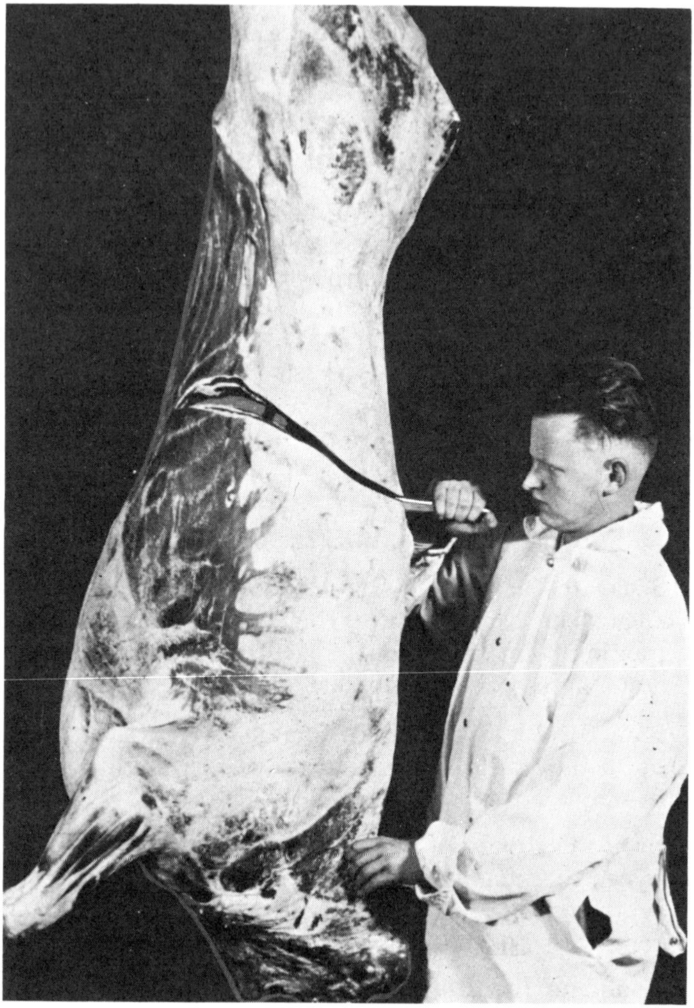

Courtesy, Morton Salt

Fig. 8.5—Illustrating the first step in quartering the carcass.

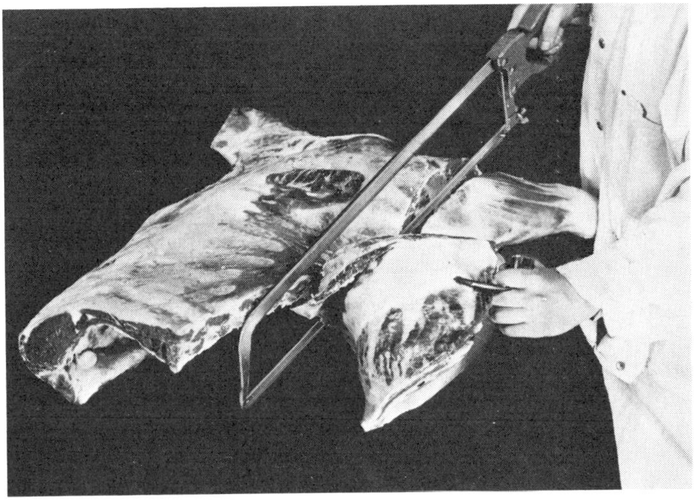

Courtesy, Morton Salt

Fig. 8.6

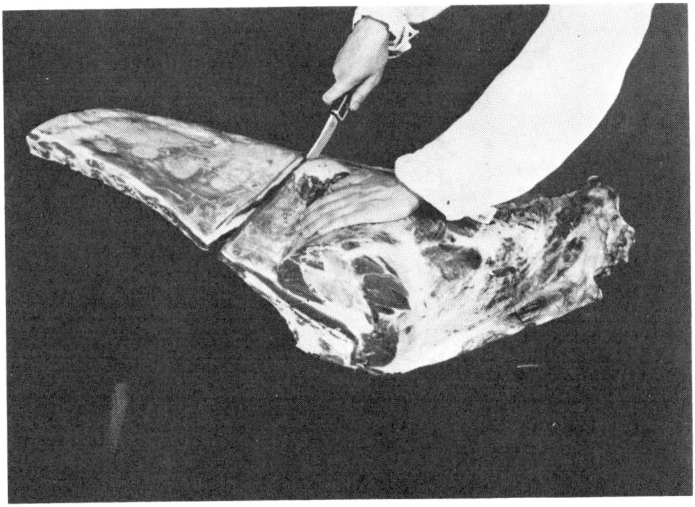

Courtesy, Morton Salt

Fig. 8.7

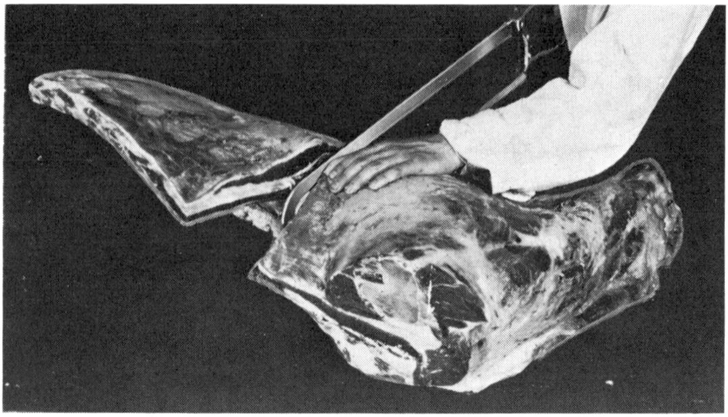

Courtesy, Morton Salt

Fig. 8.8—Complete the rib cut using a saw.

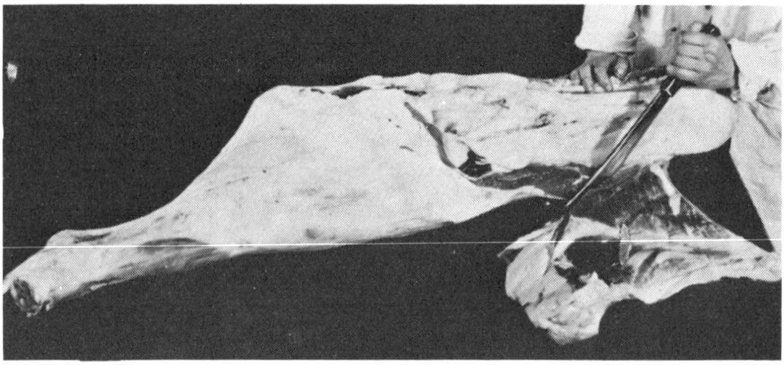

Courtesy, Morton Salt

Fig. 8.9—Removing the three-cornered flank piece from the hindquarter.

rolled if desired. The chuck is somewhat less tender and better adapted for pot roasts, corning, or grinding.

Cutting Up the Hindquarter

Remove Flank from Hindquarter—The thin three-cornered flank piece is trimmed from the hindquarter.

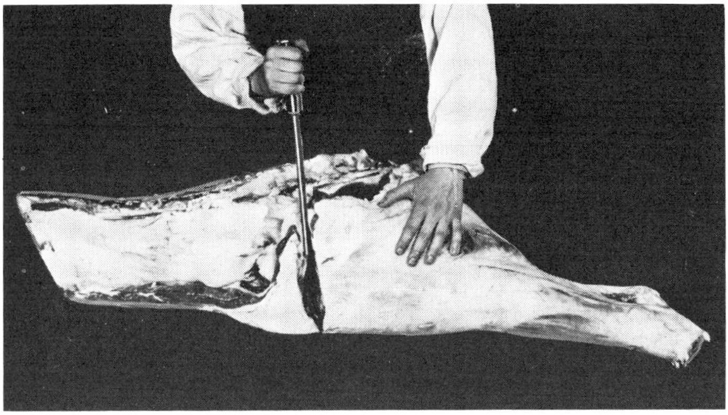

Courtesy, Morton Salt

Fig. 8.10—Removing the loin from the round.

This contains the flank steak which can be pulled out. The balance of the flank can be used for boiling piece or ground for hamburger, meat loaf, etc.

Remove Kidney—Next, trim out the kidney and kidney knob from the hindquarter, leaving sufficient fat on the backbone to cover the tenderloin.

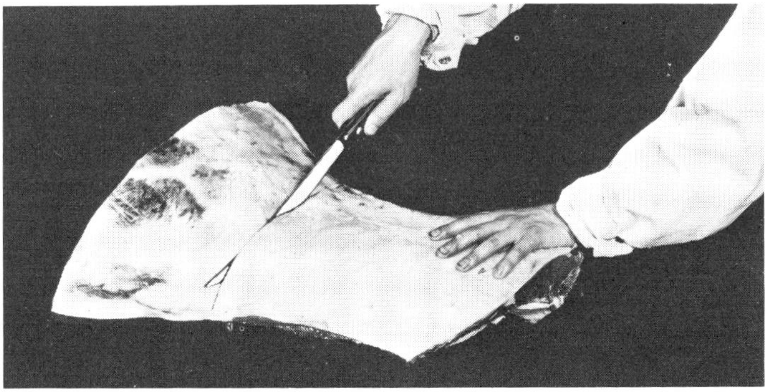

Courtesy, Morton Salt

Fig. 8.11—Separating the loin into short loin and loin end. The short loin is the one on the right.

Separate Loin from Round—The loin should be separated from the round at the ball and socket hip joint. After starting the cut with a knife, complete it with the saw.

Separate the Loin—The loin should be divided into short loin and loin end, the division being made at the hip joint. Steaks from the short loin are very popular.

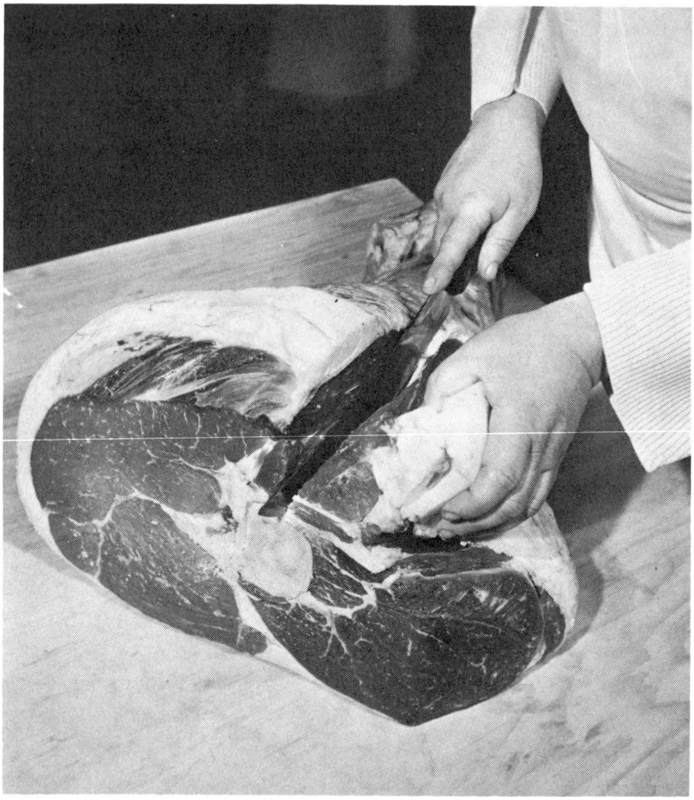

Courtesy, Swift & Co., Chicago, Ill.

Fig. 8.12—Separating the rump from the round. A saw is necessary for this operation.

Remove Rump from Round—The rump is removed from the round by sawing across the floor of the aitch bone. The rump makes a desirable roast or choice corned beef.

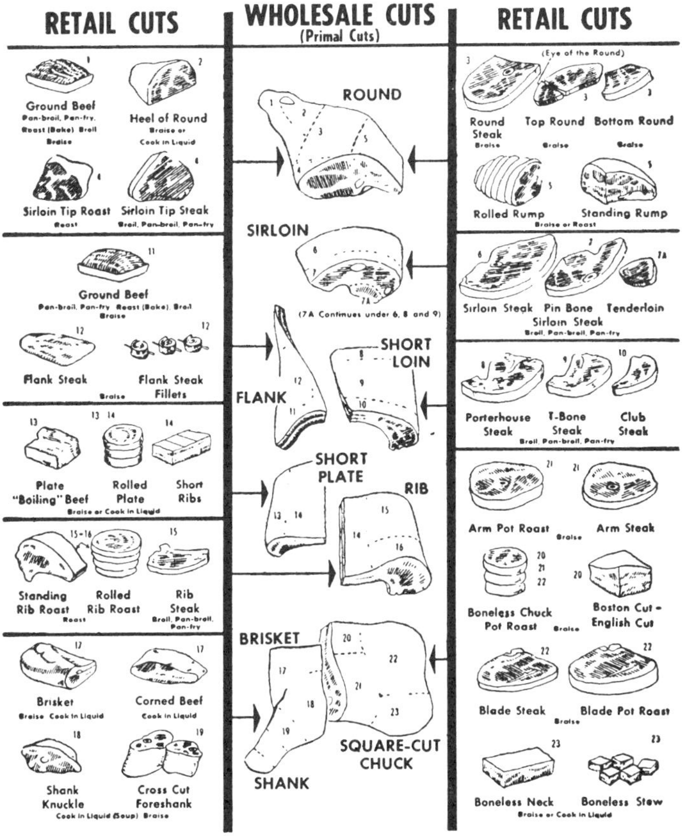

Courtesy, Swift & Co., Chicago, Ill.

Fig. 8.13—Meat packers sell most of their beef in the form of quarters—forequarters and hindquarters. To meet the specific needs of retailers, hotels and others, beef is also sold as wholesale cuts.

Utilize Cuts for Curing, Cooking, Canning

There are many different ways to use the various beef cuts in order to get the greatest value from them. Some of the cuts may be used fresh, some cured, and some canned. The picture on page 337 shows the beef carcass after making the major cuts.

Dependable refrigeration on the farm is a must in order to properly keep fresh beef. In addition, there are excellent commercial salt cures available for curing beef. Best results are obtained by following the manufacturer's recommendations for curing beef.

Handling the Hide

Sell the Hide Immediately—The best procedure if a ready market is available is to sell the hide immediately. However, if hides must be stored then the following is recommended.

Properly Handle the Hide—After carcass is skinned, remove dirt, blood, and any pieces of flesh on the hide by scraping with the back of a butcher knife and by careful cutting. Allow the hide to lose its animal heat before applying salt. When the hide has cooled sufficiently, spread it, hair side down, being sure to straighten out all folds and laps. Sprinkle fresh, clean salt over the flesh side of the hide, using about 1 pound for every pound of hide. See that all parts of the flesh side receive a sprinkling of the salt. Be sure to use plenty of salt and rub it in well along the cut edges, head, neck, legs and wrinkles.

If hide is to be stored a short period it may be folded immediately with hair side out, otherwise it should be allowed to dry spread out flat.

Market Immediately—Occasionally a buyer is available locally that will purchase green hides. This is by far the better procedure, as it eliminates curing, a messy job, and the buyer is often better prepared to cure the hide. Furthermore, buyers will not question the condition of the hide as they often do with home-cured hides.

In spite of the fact that hides are very valuable, most of the hides of homekilled beef end up being thrown away so that they become a total loss. This is true even with expertly skinned hides. A wide variety of factors are responsible such as difficulty of brand inspection, fluctuating supply and so on. However, livestock associations should grapple with this problem and reach some kind of solution that would prevent the loss of this valuable resource.

Chapter

IX

SELECTING AND USING BEEF*

Selection Important—In purchasing most foods the housewife knows what she wants. Canned goods are purchased largely by their trade names and fruits and vegetables by their quality and freshness. However, beef is usually not selected on such a well-established basis. Purchasers realize there are many kinds of steaks and roasts, but few are able to recognize the cuts of meat they desire. According to University of Florida, Agricultural Experiment Station Bulletin 321, the following factors are important in selection.

Select Meat on Conformation—Conformation indicates the proportion of lean, fat, and bone in the carcass as well as a relation between high quality and low quality cuts. A high grade carcass has short shanks and neck, well-developed round, heavily fleshed loin, and ribs with not too large a proportion of chuck and plate.

Determine the Amount of Finish—A well-finished or

*The *Meat Identification Kit* is available from The Interstate Printers & Publishers, Inc., Danville, Illinois, illustrating in color the various retail cuts of meat. Housewives and animal science students should learn to recognize them.

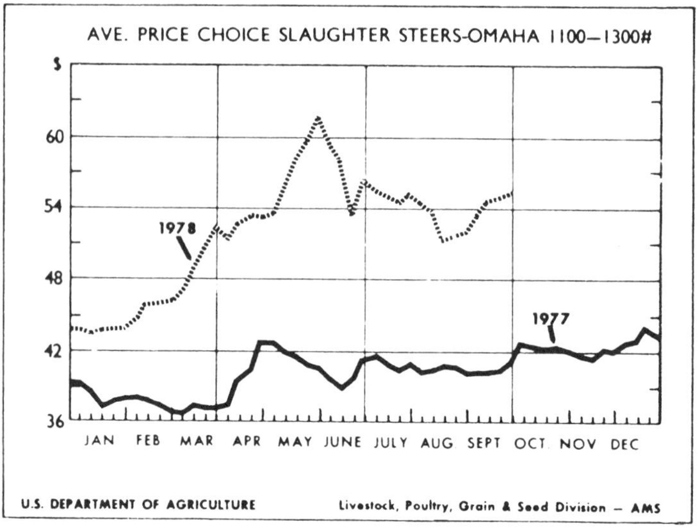

Fig. 9.0—Getting the best buy on meat is a desirable skill for every housewife. As the price of meat increases, this skill becomes more and more important. The chart indicates the wide shifts that can occur in the price of meat. Knowing how to select meat will assist one in becoming an astute buyer. A desirable practice is to watch for sales of retail beef and stock up with economical cuts of meat rather than to buy parts of a carcass.

high-grade beef carcass has an even covering of brittle, flaky, creamy white fat over the chuck, ribs, loin, and round, with a smaller amount on the shanks, neck, and plate. There are large, but not excessive, amounts of fat around the kidneys. When such a carcass is cut, relatively heavy deposits of fat are found between the larger muscles and a liberal distribution of fat occurs between the muscle bundles. This last condition, an intermixture of lean and fat, is known as "marbling."

Aging increases the degree of tenderness and improves the quality and palatability of the meat. Beef

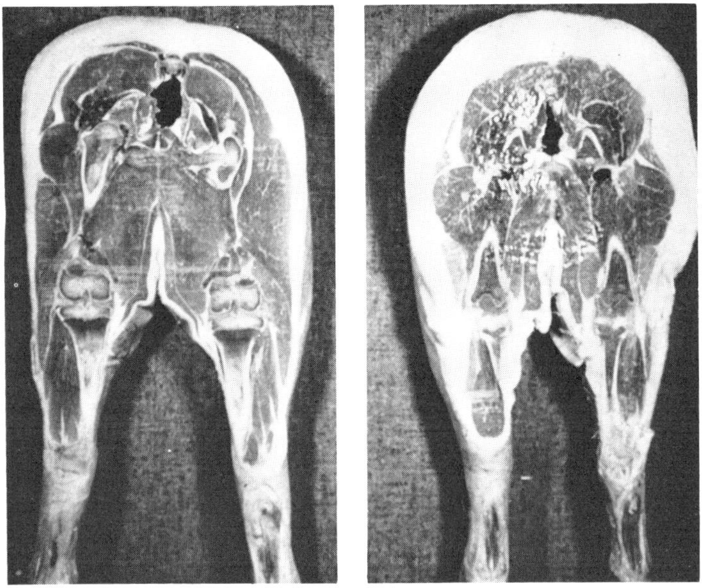

Fig. 9.1—Cross sections of two beef animals. In order to study the muscling fat covering, and relationship of live animal to carcass, both animals were frozen solid right after being killed, then each was sliced into two sections for comparison. The actual amount of red meat is of great concern in a modern breeding program. Research such as this will determine which practices livestock men will follow in the future.

with no fat covering cannot be aged or ripened sufficiently without incurring excessive trimming.

Purchase Meat by Quality—Quality refers to the degree of fineness in bones and flesh. In good quality beef, the bones are relatively small and if from a young animal are red in color and porous in character on the cut surfaces. The lean is fine in texture, bright red in color, smooth and velvety in appearance, and well marbled with creamy, white, flaky fat.

Grade Is Important—The U.S.D.A. and various meat

packing companies have established certain grades for fresh beef, and a uniform, definite standard for each of these grades is gradually becoming recognized.

Quality Grading

Established in the 1920's, quality grading is the oldest system of beef grading, and theoretically rates carcasses according to the palatability traits of tenderness, juiciness, and flavor.

The Quality grades are Prime, Choice, Good, Standard, Commercial, Utility, Cutter, and Canner. Quality grade is generally reported to the nearest one-third of a grade—such as Choice minus, Choice, and Choice plus.

The most important factors in determining palatability of beef are maturity and marbling, although the USDA grader evaluates five different traits before assigning a Quality grade to a carcass.

1. Maturity

The older the animal, the tougher the meat. The grader determines an animal's age by examining a cross-section of the spinal column and noting the degree to which cartilage has ossified (hardened to bone).

Most fed cattle fall into the "A" maturity bracket—9 to 30 months of age. To be eligible for Prime, Choice, Good, or Standard grades, cattle must be either "A" or "B" maturity—that is, less than 42 months of age.

2. Marbling

Marbling refers to the amount of fat within the muscle tissue of a carcass. There are nine different grees of marbling, ranging from "practically devoid" bundant." Marbling has been thought to contribute greatly to the eating satisfaction of beef, although recent USDA research indicates that marbling levels above "slight" do not significantly raise palatability factors in animals of "A" maturity. The USDA grader arrives at a marbling score by looking at a ribeye cross-section of the carcass.

3. Color of Lean

Color of beef muscle tissue is important because it has considerable influence on consumer acceptance. If it is abnormally dark or light, Quality grade can be discounted as much as one and one-half grades. The desired color of lean in beef is described as "cherry red."

4. Texture of Lean

Lean muscle has a "grain" or texture just as wood has. The preferred texture is fine and smooth. Coarse texture is associated with older animals and discounted in grade.

5. Firmness of Lean

Firmness is an important quality factor since soft, watery meat is unacceptable to the consumer. Problems with color, texture, and firmness of lean are not widespread, so maturity and marbling are the major factors considered in quality grading.

Yield Grading

Yield grading was begun by USDA in 1965. It estimates the quantity of lean, edible meat that can be cut from a carcass. Carcasses are rated from 1 (highest yield and most desirable) to 5 (lowest yield and least desirable). Yield grade is generally reported to the nearest one-tenth of a grade.

The four traits used to determine Yield grade measure the proportion of lean meat to fat in the carcass:

1. Fat Thickness

This is a measure of the thickness of fat covering the outside of the carcass. It is the most important of the four traits determining Yield grade, and is used to establish a "preliminary Yield grade," which is then adjusted to the grader's findings in the other three traits.

Fat Thickness	Preliminary Yield Grade
0.00 inch	2.0
0.20 inch	2.5
0.40 inch	3.0
0.60 inch	3.5
0.80 inch	4.0
1.00 inch	4.5
1.20 inches	5.0
1.40 inches	5.5

The USDA grader uses a measuring device to determine fat thickness of the carcass. The measure is taken over the ribeye muscle at the twelfth rib, and is sometimes adjusted by the grader to a figure that he feels is more representative of the outside fat on the entire carcass.

2. Hot Carcass Weight and Ribeye Area

Hot carcass weight and ribeye area are adjusted for simultaneously. As a beef animal gains weight, it becomes more heavily muscled. So as hot carcass weight increases, the grader should observe a corresponding increase in ribeye size. The larger the ribeye, the greater the yield of retail product, and the lower the Yield grade number will be.

The USDA grader uses a grid to measure ribeye area. If the ribeye is smaller or larger than average for the animal's carcass weight, he adjusts the Yield grade number up or down. For example, a 1-inch deviation from average ribeye size means an adjustment in preliminary Yield grade of + or − 0.3 of a grade.

3. Percent Kidney, Heart, and Pelvic Fat

This is a measure of the amount of fat found within the internal cavities of the carcass. The bulk of this internal fat is found adjacent to the kidney. The less internal fat, the more desirable the Yield grade.

The USDA grader notes the amount of internal fat in a carcass. The average beef carcass has 3.5% of its total weight in the form of kidney, heart, and pelvic fat. For every 0.5% deviation from the average, the preliminary Yield grade is adjusted + or − 0.1.

Select Meat for Flavor—Flavor in meat is determined largely by the presence or absence of meat extractives and internal and external fat. As the animal matures, meat extractives increase in the muscle. Therefore, meat from a mature animal may lack somewhat in tenderness but has sufficient meat extractives to give a well-flavored product.

Beef from heifers 16 to 30 months of age of good conformation and a moderate amount of finish should be well flavored and comparatively tender.

Select the Proper Cut of Meat

Select Suitable Size Cuts—One objective in cutting meat is to obtain pieces suitable for use in the home.

Round Steak—Round steak, the least expensive

steak of the hindquarter, is oval in shape with a small round bone in the center. It consists of several large muscles but lacks in exterior and interior fat. The cut shown, however, has more fat throughout the lean tissue than is found generally in round steak. The steak in the illustration is from the center of the wholesale round which is higher in quality than steaks cut from either the loin or shank end. The shank end of the round contains a relatively small amount of lean meat and a large amount of bone and connective tissue. It is used as pot roast, stew, or ground meat.

Rump Roast

The rump is a wedge-shaped piece of meat lying between the round and loin end. It has a moderate amount of fat and a relatively large portion of bone, as it contains the hip joint and part of the pelvic arch. This cut is suitable for roasting, and for this purpose

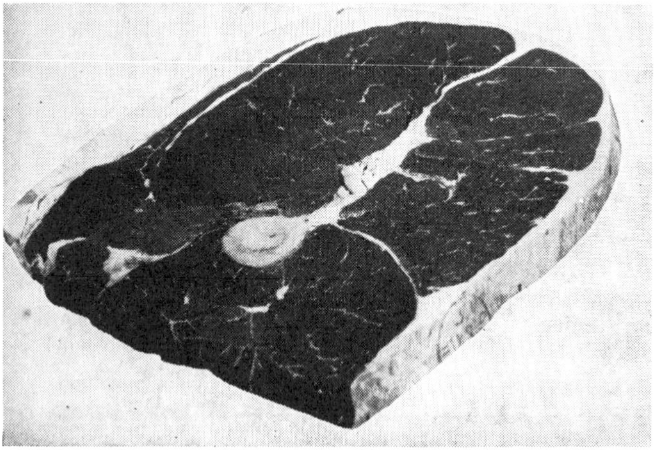

Fig. 9.2—Round steak, the least expensive steak of the hindquarter.

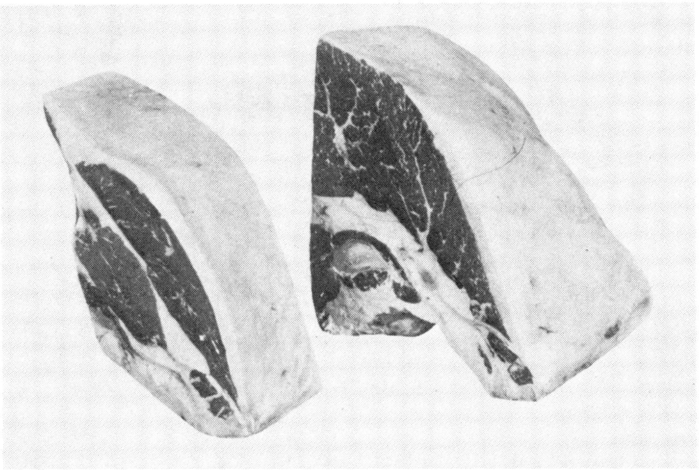

Fig. 9.3—Rump roast.

it frequently is boned and rolled. The lean portion can be used as boneless steaks for broiling or pan frying.

Round Bone Sirloin Steak

Sirloin steaks are obtained from the wholesale loin end cut and are characterized by being oblong in shape and containing sections of the backbone and hip bone. No two sirloin steaks are the same in muscle and bone because of the changing positions of these structures. The steaks procured from the loin end are considered more tender and to have a more desirable flavor than those from the wholesale round cut.

Double Bone Sirloin Steak

The characteristics of this steak are muscle fibers running in two directions and a relatively large proportion of fat and bone to lean meat. When lean meat is

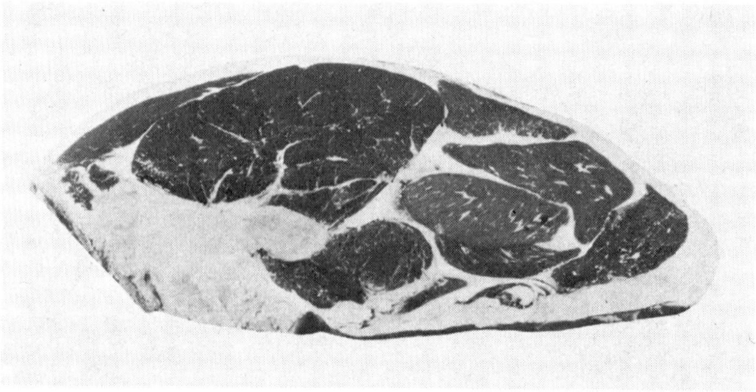

Fig. 9.4—Round bone sirloin steak.

cut parallel to the muscle fibers (see right of figure) it lacks tenderness, and for this reason, it is not as desirable as the round bone sirloin steak. This steak may be broiled and when cut three inches thick makes an excellent roast.

Porterhouse Steak

Porterhouse, T-bone, and club steaks are obtained

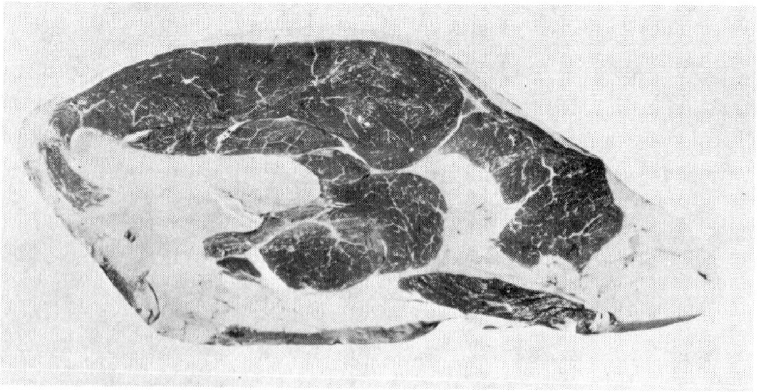

Fig. 9.5—Double bone sirloin steak.

from the short loin. Porterhouse is considered to be the highest quality steak in the beef carcass. It is characterized by the T-bone conformation, large "eye of beef" above the bone, and the tenderloin muscle below. When the tail end of the steak, which is rather coarse, is cut off, the resulting steak is known as a "short cut" porterhouse. These steaks should be cut about ¾ inch thick if they are to be cooked well done, but if slightly rare meat is preferred, 1 to 1½ inches is not too thick. Porterhouse, because of its flavor and tenderness, is in greatest demand and, therefore, is the most expensive cut in the beef carcass.

T-Bone Steak

T-bone steak can be recognized readily by the distinct T-bone formation. This steak shows a considerable

Fig. 9.6—Porterhouse is considered the highest quality steak in the beef carcass although seldom offered as a choice in restaurants.

Fig. 9.7—T-bone is similar in quality to porterhouse.

portion of tenderloin muscle, but usually there is much less than in porterhouse steak, or none at all. The "eye of beef" above the bone is similar in quality to that of the porterhouse. This cut, as indicated in the figure, contains a considerable quantity of kidney fat or suet. It is cut and cooked as is porterhouse and the price is approximately the same.

Club Steak

One or two club steaks are obtained from the rib end of the short loin. These steaks have neither T-bone nor tenderloin muscle, but may or may not have a rib. Club steaks have a large "eye of beef" which in good quality beef is well marbled and surrounded on the exterior

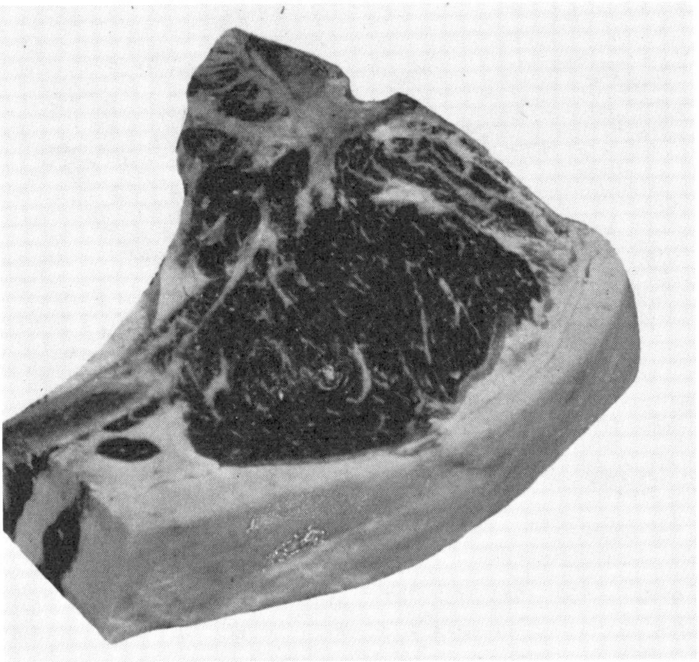

Fig. 9.8—Club steak is a high quality cut.

by fat. This is a high quality cut and usually is broiled. For oven broiling, 1 to 1½ inches in thickness is best. Club steaks are considered to be slightly inferior in quality to porterhouse and T-bone steaks. However, they have a higher percentage of lean to fat and bone than do the other steaks obtained from the short loin.

Flank Steak

The flank is the cheapest wholesale cut of the hindquarter. It contains about 50 per cent surplus fat and connective tissue, and very little bone. Its chief value lies in the large abdominal, flat, rectangular-shaped muscle which is known as the flank steak. The muscle

Fig. 9.9—Flank steak of medium tenderness but rich in flavor.

fibers of the steak are large and run lengthwise. When used for Swiss steak, it is scored on both sides diagonally to the fibers. When rolled with suet or fat pork, it may be used as a roast. This steak is rich in flavor and when properly prepared is of medium tenderness. It is best when cooked with moist heat in a low temperature oven for a long time.

Chuck Rib Roast

The square cut chuck has many small muscles and as the muscle bundles run in various directions, it is impossible to cut all the muscles across the grain. The chuck has a relatively large amount of connective tissue, little fat, and a fairly high percentage of bone. A chuck rib roast from a carcass which grades "medium" or higher makes a satisfactory roast or it may be cut into chuck steaks. It is well flavored because it contains a relatively large amount of meat extractives.

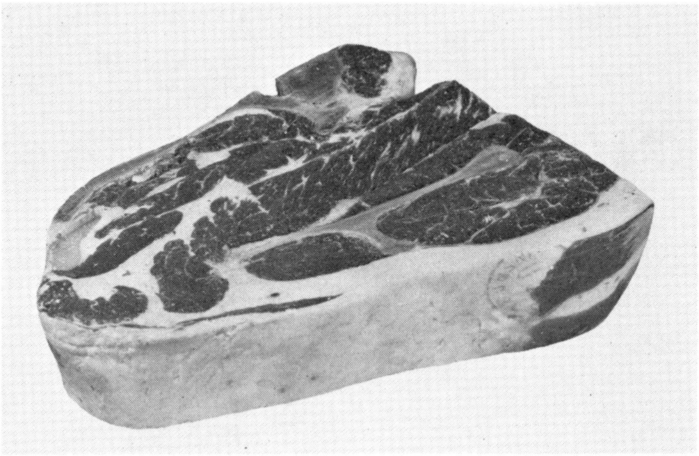

Fig. 9.10—Chuck rib roast, which is well flavored and generally an economical buy. Stock up during sales and store in the home freezer.

Boned and Rolled Chuck Rib Roast

The chuck rib roast when taken from a medium-sized carcass may be too large for a small family. This roast is boned and made into two rolls. Each of these rolls weighs approximately 4 pounds. The rolls contain no bone, a good proportion of lean and fat which is well distributed, and very little connective tissue.

Round Bone Pot Roast

Round bone pot roast is cut across the ribs and at right angles to the chuck roasts. It is known also as cross rib roast or arm bone roast because it is cut across the ribs and has in the center, the small round arm bone. It may be cut into steaks for pan frying, and when cut of suitable thickness it is used for Swiss steak. When cut into steaks it sometimes is sold as round steak.

Fig. 9.11—Boned and rolled chuck rib roast.

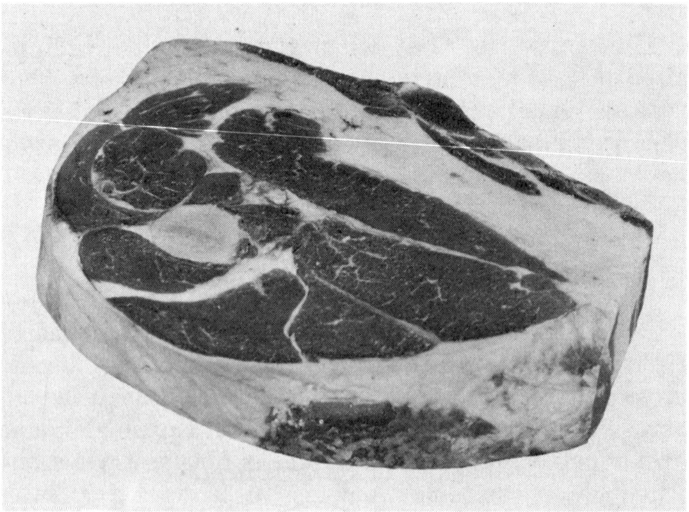

Fig. 9.12—Round bone pot roast, also known as cross rib roast or arm bone roast.

Fig. 9.13—Boneless neck.

Boneless Neck

The neck end of chuck shows a large proportion of lean meat. However, this piece contains the large shoulder knuckle, one rib, and all of the neck vertebrae. In good quality carcasses, one or two pot roasts can be cut from the thicker part. These will be similar in quality and texture to the chuck roast. The neck proper is well flavored meat but lacks tenderness. Boneless neck, because of its flavor, is one of the most desirable cuts for ground beef. Chopping or grinding finely divides the connective tissue present and thus the meat is made tender. It also makes excellent stew meat.

Shoulder or Soup Knuckle

The shoulder or soup knuckle contains a large amount of bone, connective tissue, and varying amounts

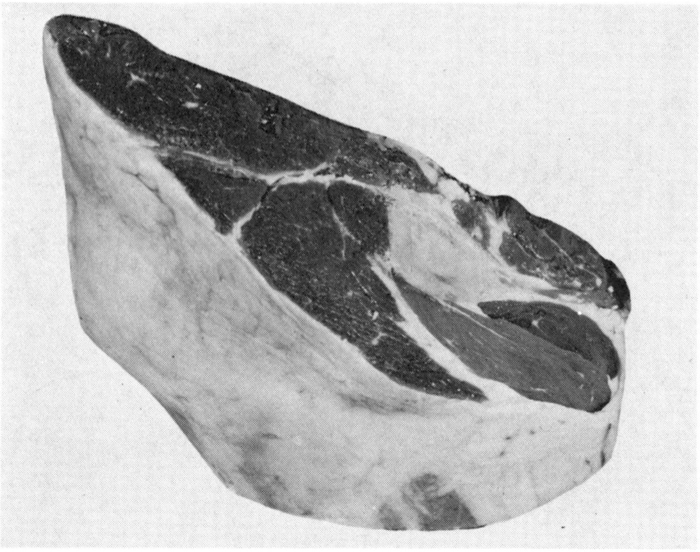

Fig. 9.14—Shoulder or soup knuckle.

of lean meat, depending upon the method of cutting. This cut, as the name implies, is used for soup.

Fore Shank

The fore shank has a large amount of bone and connective tissue. The connective tissue makes it a highly desirable cut for soup or for pressed or jellied beef. A small pot roast or several shoulder steaks may be cut from the large or upper end. The lower part of the shank is cut into 2- or 3-inch lengths to use for soup, or the meat may be used for stew or ground beef.

Brisket

The plate of beef consists of the navel and brisket.

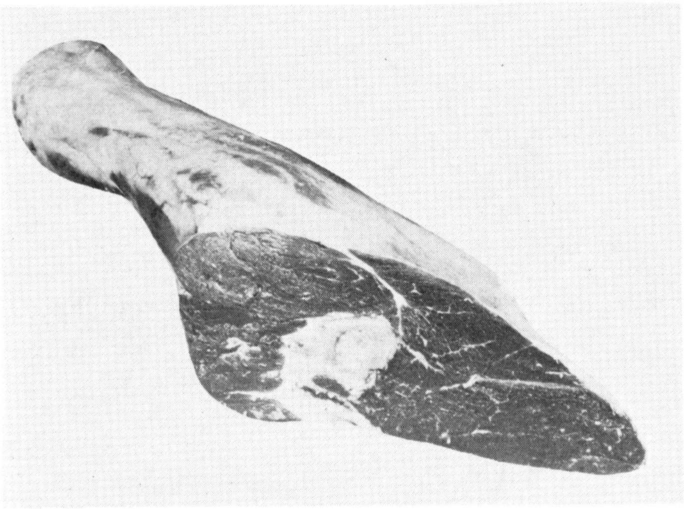

Fig. 9.15—The fore shank.

The lean and fat, which are well flavored, are deposited in alternate layers. The plate contains the lower end of the ribs and breast bone. The brisket can be identified from the navel by the presence of the breast bone and the granular nature of much of the exterior fat. There is a large amount of bone and fat, and the fat when cooked has a sweet flavor. This cut is used for soup, stews, and ground beef.

Boiling Pieces from Plate

Sections consisting of two or more rib ends are cut from the plate and sold as boiling beef. Note the alternate layers of lean and fat in the folded cut. These cuts usually have moderate amounts of fat and the lean is well flavored.

Fig. 9.16—The brisket.

Short Ribs of Beef

Short ribs of beef are cut from the upper portion of the navel. They are cut from 2 to 4 inches in length and are used for boiling or as a pot roast. The short ribs are considered to be the best part of the plate.

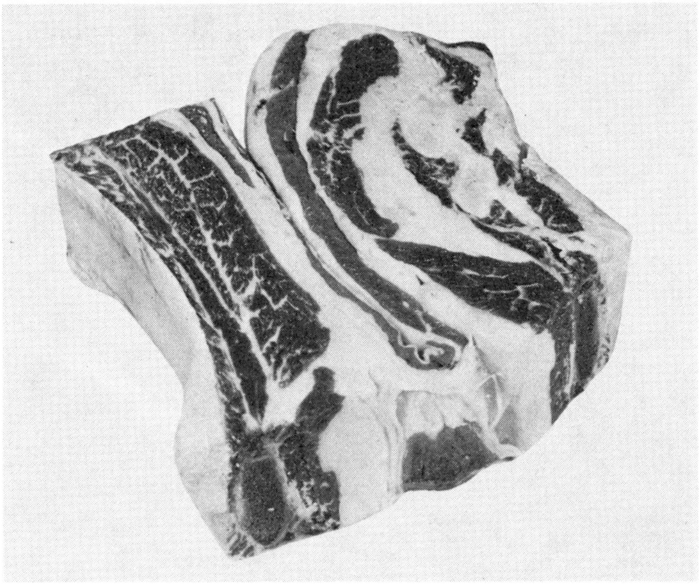

Fig. 9.17—Boiling pieces from plate.

Fig. 9.18—Short ribs of beef, used for boiling or as a pot roast.

Fig. 9.19—Boned and rolled navel.

Boned and Rolled Navel

The navel piece can be boned, rolled, and cut into pieces of any desired size. Note the thin outside lean portion which is fibrous and lacking in tenderness. The inside of the roll contains desirable proportions of lean to fat and is well flavored. If the roll contains a moderate amount of fat, it can be used more frequently for boiling or stewing than for any other purpose.

Recipe for Beef Jerky

1 lean round of beef from mature animal. Cut meat into long thin chunks or strips about 1 to 1½ inches wide and not over ½ inch thick. If possible, cut across the grain. Lay strips in large pans and sprinkle each layer with jerky mix.

Jerky Mix:

3 parts salt
1 part ground black pepper
1 part garlic salt or powder
1 part seasoned salt (any brand)

Mix well, then allow beef strips to set in the seasonings for 3 or 4 hours. Hang strips over tightly stretched cotton clothesline. (Large cotton rope is best.) Hang in a shady place protected from flies if possible. About four or five days will dry the meat sufficiently for storage in a loose weave sack. Approximately 7 pounds of fresh meat will make 1 pound of jerky.

Recipe for Spiced Tongue

1 fresh beef tongue—well scrubbed.

Water to cover
½ c. cider vinegar
2 bay leaves
2 cloves of garlic
1 tsp. mustard seed
1 tsp. whole cloves
1 tsp. celery seed
1 Tbsp. salt
1 tsp. pepper
2 Tbsp. sugar

Put tongue in a large kettle. Cover with water—just level. Add all spices and seasonings, bring to a boil, then turn heat to simmer and cook till tender (about 2 to 3 hours). Test with fork. Set aside and cool in broth. Then peel off outer skin and trim off any excess fatty tissue. Chill and slice. Serve cold for sandwiches.

Chapter X

MARKETING BEEF

There are two ways in which to increase the profit in producing beef cattle. First is to lower the cost of production and the other is to increase the selling price. Therefore, ranchers need to give special consideration to selecting a suitable market, time of year to market, and numerous other factors in marketing beef. There are a number of approved practices which should be followed in marketing for the greatest possible profit.

Activities Which Involve Approved Practices

1. Selecting a suitable market.
2. Marketing beef at a desirable time.
3. Selecting the proper age to sell.
4. Properly sorting cattle.
5. Handling cattle properly.
6. Preventing shipping losses.
7. Preparing cattle for sale.
8. Fitting cattle for exhibition.
9. Photographing beef cattle.

1. Selecting a Suitable Market

Cattle may be marketed by a number of marketing systems. The cattle producer should make a careful

Courtesy, Union Pacific Railroad

Fig. 10.0—The Stockyards at Ogden, Utah.

study of the different methods before deciding upon the one best suited to his individual needs. Different markets in certain parts of the country will pay a premium price for cattle best suited to their tastes and needs.

Public Stockyards—The great bulk of cattle are assembled and sold through the medium of the public stockyards. When cattle are shipped to the stockyards, they are assigned to a livestock commission firm. The livestock yards feed and water the cattle and pen them so they can rest and improve their appearance. The commission firm generally sorts the cattle into market groups as desired by buyers. After the cattle are sold, the consignee deducts from the proceeds the marketing costs, such as freight, insurance, yardage, feed, and commission fee, and pays the balance to the consignor. One advantage of this method of selling is that any number and all types of cattle may be sold at any time of year. Commission firms generally keep the customers informed as to market conditions.

Cooperative livestock sales agencies on public markets are producer-owned commission firms. The agen-

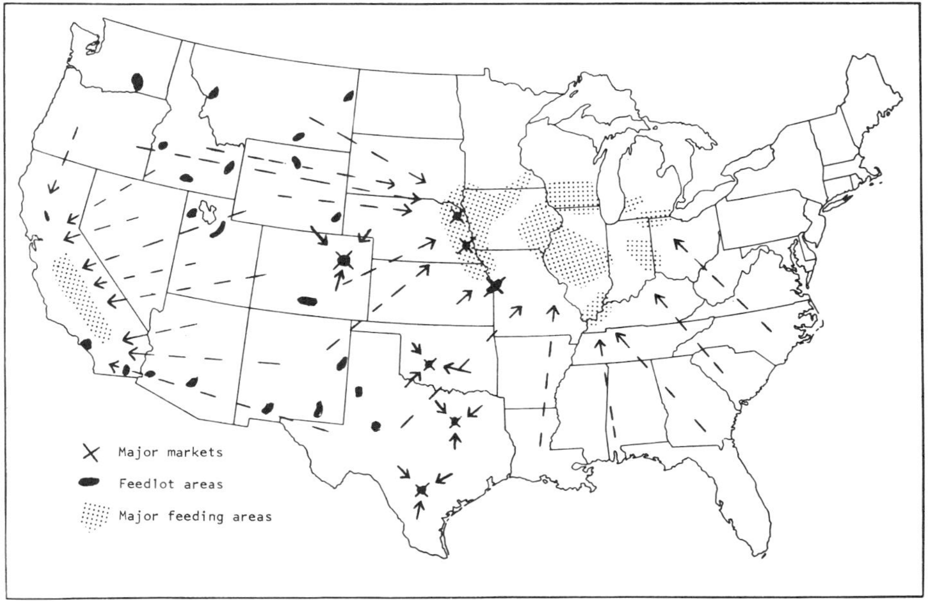

Courtesy, California Agricultural Experiment Station

Fig. 10.1—Movement of stocker and feeder cattle to major markets and feeding areas. Understanding the movement of cattle is a desirable practice in improving the marketing of one's own livestock.

cies charge the commissions allowed at the market, but distribute the surplus earnings to their members in the form of patronage dividends.

Direct Markets—Cattle may be marketed direct through such agencies as local dealers, packing plants, local meat markets, and auction sales. Local dealers usually have their own trucks and go to farms to get livestock. They then transport it to a point where they desire to sell. Many cattle are handled in this fashion. Frequently the price is lower than at competitive markets; however, it must be remembered that a service

Fig. 10.2—Periodically weigh cattle or representative groups of cattle, so weight and market price can be correlated. A good set of scales is a desirable piece of equipment on any livestock ranch.

is performed in assembling and transporting cattle to distant points.

Packing plants slaughter large numbers of cattle and calves each year. The rancher can market direct to these plants and know the price he will receive in advance. Cattle are generally weighed and sold upon arrival. Representatives of the various packing plants travel throughout the cattle country buying cattle. These men are well acquainted with market conditions in various parts of the country.

Local meat markets often afford a desirable market in many smaller cities. Farmers who have only a few animals to sell often can obtain a better price through these markets, particularly if they have the type of cattle that local conditions demand.

Fig. 10.3—An assembly line technique used to groom animals for auction. Cattle must look their best before entering the sales ring. Here each person has a special task to perform as the animals move from station to station. Full advantage is taken of power equipment such as electric blowers, clippers, etc., in order to speed up the process as well as to do it better.

Fig. 10.4—Local auction sales are a popular method of buying and selling livestock.

Auction sales are well known and becoming increasingly common in many parts of the country. Cattle are brought to the sales center by the owner and auctioned to the highest bidder. This is frequently a good method to sell underweight cattle, as buyers with a surplus of feed often obtain additional cattle from this source. The auctioneer charges a fee or commission for selling livestock in this manner.

New Markets—Ranchers should constantly be on the lookout for new ways of selling livestock in order to get the highest possible return. If your cattle are better than average, you might sell replacements at a premium to your neighbors.

Custom feed lots enable the rancher to send his cattle to a feed lot whose specialty is feeding and fattening cattle. By putting his cattle in the custom feed lot, the farmer can protect against having to sell during heavy market runs, can hold and fatten for a more opportune time.

Selling Purebred Cattle—A breeder of high quality cattle must develop his own market and create a demand for his cattle. There is a large field here, as many farms need higher quality livestock. Increased interest in crossbreeding increases the demand for purebreds.

Private treaty. Most purebred cattle are sold by private treaty. Buyers are attracted by the excellence of the livestock and the reputation of the breeder.

Consignment sales. Auctions wherein several breeders consign cattle are called consignment sales.

Breeder auctions. If an individual breeder has 35 or more head to sell, he often holds an auction on his own farm. Proper facilities are necessary, such as a pavilion, if this method is to succeed.

Mail order. If the breeder has an excellent reputation, many buyers prefer to trust his judgment and save themselves the expense and time involved in personal inspection.

Sell on grade and yield. This is a method of buying cattle "on the rail" rather than live. As yet only a small percentage of cattle is sold this way, mostly in the southwest. U.S.D.A. estimates that less than 5 per cent of all U.S. cattle are sold on a grade-and-yield basis. It involves a method whereby cattle are slaughtered and the packer grades the carcass and thereby establishes the price. This method eliminates to a large degree the competitive bidding aspect of cattle sales, but does offer one method of selling livestock based on quality of carcass and yield of meat.

2. Marketing Beef at a Desirable Time

Study Demand—Worldwide there are great differences in the meat-eating habits of people. In the United States, steer beef commands a higher price than heifers, whereas in Canada and England, the opposite is true. Jewish people prefer the forequarters (kosher beef); however, meat from the rear quarters generally brings a higher price in most of the United States. The German people are sausage eaters, so meat processors in that country prefer the forequarters of cattle. They import many of these from France, where the reverse is true. Historically, the French have used hindquarters and many of these buying habits are an important part of the culture of a nation and its people, regardless of where they live.

In the United States male animals are castrated and sold as steers because they bring a higher price. In Germany, for example, approximately 90 per cent of male cattle are sold as bulls because bulls are more efficient to feed. There are associated problems with the practice of selling bulls but it is a trend that will probably grow in the United States. Such wide variations do not exist within a local community; however, to a degree it is true. For example, some communities prefer veal, so this makes for a good calf market. Others prefer very lean meat, so the price of fat cattle could be affected.

CATTLE: WEEKLY AVERAGE PRICES
In dollars per 100 pounds

Slaughter Cattle		Iowa 1/ Mar. 4, 1978	Iowa 1/ Mar. 3, 1979	California 2/ Mar. 4, 1978	California 2/ Mar. 3, 1979	Texas Panhandle 2/ Mar. 4, 1978	Texas Panhandle 2/ Mar. 3, 1979	Colorado 3/ Mar. 4, 1978	Colorado 3/ Mar. 3, 1979	Washington-Oregon 2/ Mar. 4, 1978	Washington-Oregon 2/ Mar. 3, 1979
Steers -											
Prime 3-4	900-1100#	—	—	—	—	—	—	—	—	—	—
	1100-1300	—	—	—	—	—	—	—	—	—	—
Choice 2-4	900-1100#	45.35	67.32	48.25	70.75	45.82	68.12	45.62	67.10	46.50	68.75
	1100-1300	46.48	68.05	48.00	70.00	45.82	68.12	45.62	67.10	46.50	68.00
Good 2-3	900-1100#	41.95	61.90	47.00	70.00	—	—	—	—	—	—
Heifers -											
Choice 2-4	900-1100#	44.08	66.00	44.50	67.00	43.50	65.50	43.55	65.45	44.25	65.75
Good 2-3	700-900	39.95	59.90	—	—	—	—	—	—	—	—

Feeder Cattle		Georgia Auctions Mar. 4, 1978	Georgia Auctions Mar. 3, 1979	California 4/ Mar. 4, 1978	California 4/ Mar. 3, 1979	Kentucky Auctions Mar. 4, 1978	Kentucky Auctions Mar. 3, 1979	Colorado-Kansas Auctions Mar. 4, 1978	Colorado-Kansas Auctions Mar. 3, 1979	Illinois Direct Mar. 4, 1978	Illinois Direct Mar. 3, 1979
Steers -											
Prime	300-400#	—	—	—	—	—	—	60.03	116.25	—	—
	400-500	—	—	—	—	—	—	57.67	107.16	55.20	93.60
	500-600	—	—	—	—	—	—	54.50	97.29	53.70	90.05
	600-700	—	—	—	—	—	—	50.47	87.21	50.45	84.15
Choice	300-400#	55.00	117.50	—	—	52.75	108.50	57.59	99.97	—	—
	400-500	51.50	102.50	56.00	94.50	51.50	101.00	55.85	100.28	49.95	85.25
	500-600	48.00	90.50	54.75	87.00	48.75	87.50	52.50	92.49	48.20	82.75
	600-700	45.50	81.00	50.50	81.00	47.60	83.50	49.59	84.80	47.20	80.65
	700-800	44.50	77.00	48.25	77.00	45.25	79.50	47.47	80.52	46.70	75.40
Good	300-400#	48.50	105.00	—	101.00	44.00	92.00	54.93	102.00	—	—
	400-500	46.50	92.50	—	96.00	43.75	84.00	53.50	96.62	—	—
	500-600	44.00	81.50	—	—	42.50	79.00	50.73	88.62	—	—
	600-700	41.50	72.50	—	73.00	41.00	76.00	48.35	79.56	—	—
Standard	300-400#	40.00	87.50	—	—	35.50	82.00	—	—	—	—
	400-500	39.00	78.00	—	—	35.50	80.50	—	—	—	—
	500-600	38.50	71.50	—	—	35.25	74.50	—	—	—	—
	600-700	37.00	67.00	—	—	34.50	72.50	—	—	—	—
Heifers -											
Choice	300-400#	43.00	92.50	45.75	83.00	41.25	89.50	48.76	94.89	—	—
	400-500	41.00	83.50	45.38	80.50	41.50	82.50	47.28	90.08	44.70	78.80
	500-600	41.00	75.00	44.25	75.00	42.00	79.50	45.88	80.68	44.05	75.10
	600-700	—	—	43.38	69.50	40.00	74.00	44.92	74.75	42.95	72.40
Good	300-400#	39.50	82.00	—	—	36.00	80.00	46.20	91.58	—	—
	400-500	38.00	77.00	—	—	36.75	76.50	45.23	84.67	—	—
	500-600	38.00	71.50	—	—	37.00	74.50	44.00	77.17	—	—
	600-700	—	—	—	—	36.50	73.00	43.93	72.13	—	—

1/ Most cattle delivered to plant with weighing conditions equivalent to 3% shrink at feedlot.
2/ F.O.B. feedlot 4% shrink.
3/ Weighed feedlot 4% shrink and delivered plant.
4/ 3% shrink or equivalent.

Fig. 10.5—Study market reports. Above is one from the USDA *Livestock Market News*. Marketing agencies will send them at no charge. Livestock magazines and newspapers also feature this information.

At any rate, it is a desirable practice to study all phases of marketing beef in order to squeeze out the most profit for a producer's efforts.

Study Markets—When to sell is always a problem, but frequently the early market is the best. In range country, cattle make most of their gains on summer grass by August. Extra feeding will top out the herd. This will leave more grass for the rest of the cattle and also protect the range.

Market Early—Cattle marketed early are generally in greater demand and consequently sell at a higher price than when supply is heavy. In mild climates some cattle feeders make a practice of purchasing cows and late calves in the fall, feed them through the winter, and dispose of all animals by late spring or early sum-

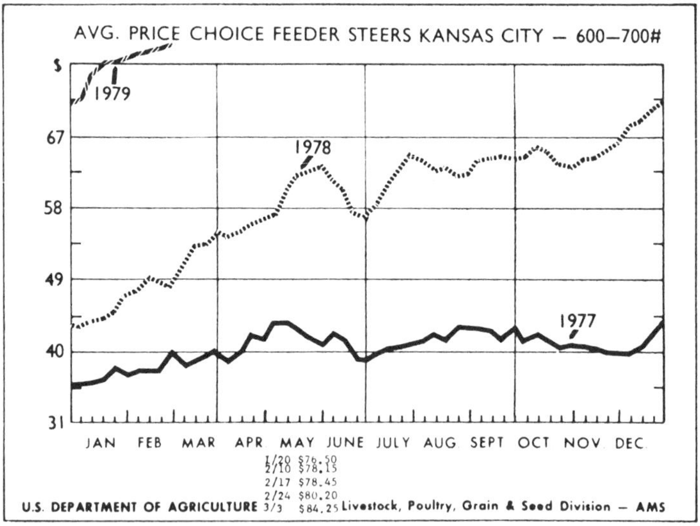

Fig. 10.6—This chart indicates yearly trends. Livestock producers can use this knowledge to advantage in either buying or selling animals.

mer. This leaves the range in excellent shape with a good supply of feed available for new feeders purchased in the fall. This factor is not so important with fed cattle, as location of market and degree of finish receive most attention.

Determine Demand in Advance—It is a desirable practice to contact dealers or commission firms before cattle leave the ranch. Occasionally, holding cattle a day or two will make a considerable difference in selling price. Religious holidays, total supplies, and other factors often affect the market. The producer should know as much about these as possible, but experienced commission men should also be consulted.

Ship Cattle When Ready—This rule cannot be followed blindly; however, there is an extended period when cattle are in proper condition and every effort should be made to dispose of them during this state. Cattle are difficult to keep in peak condition, and gains become more expensive. However, cattle should never be sold with no regard to market conditions even should they be ready to go.

Use a Contract—Livestock producers could do much toward increasing marketing efficiency through use of simple yet sound purchase or sale contracts.

Many contracts now in use fail to meet all of the essentials of a good contract. Actually, a number of so-called "contracts" are not contracts at all but mere "options to buy." Inequities and tremendous losses are suffered each year by both producers and buyers because of faulty contracts or failure to use written contracts. Most county agents can supply information on reliable contracts.

3. Selecting the Proper Age to Sell

Study Cost Records—Over a period of years, cost

Fig. 10.7—A No-Bak set of gates installed in a chute. These swinging gates allow stock to go through but prevent their backing up, thereby speeding up the movement of cattle through the chute.

records for the individual ranch coupled with information on market demand are the best basis for deciding at what age to market and whether to finish the cattle or sell them as feeders.

Consider Amount of Beef Produced from Given Amount of Feed—In general, cost of production records on breeding ranches show that the total pounds of beef produced from a given amount of feed are greatest when cattle are sold as yearlings.

Total ranch production in the form of two-year-old feeders is usually less than if all the surplus is sold as yearlings and a larger breeding herd is kept.

4. Properly Sorting Cattle

Sort Cattle into Uniform Groups—The buyer who is seeking a uniform lot of top quality cattle will discount his bid if inferior animals are included in the offering. On the other hand, the buyer who is satisfied

Fig. 10.8—Cattle sorted into a uniform group before marketing.

with plain cattle may pay the full market price for that kind, but will pay little if any premium for top quality.

Sort According to Quality and Weight—The wider the variation in quality and weight, the greater the need for sorting. Practical sorting of animals before offering for sale should be part of the policy whether the animals are to be sold at home or elsewhere.

5. Handling Cattle Properly

It is important to properly handle cattle in order to avoid injury to livestock and insure efficiency of labor and operations. It is doubly important to properly handle cattle that are being sold in order that they will be in the best possible condition for the buyer. Following are some suggestions for handling cattle:

a. Do not "rush" cattle. Let them see where they are going.
b. Drive the entire herd into an enclosure rather than trying to separate one or two animals away from the group.
c. Keep calves away from their dams until they are hungry, then watch them "pair up" in order to establish identity of calf.
d. Cattle will always walk the fence line in a new pasture. Be certain there are no weak spots which they can go through.
e. Avoid injury and possible later infection caused from kicking, beating, using the pitchfork on them, etc.
f. Learn to speed up slow critters by twisting their tails, sticking them with a sharp instrument like a pencil, using an electric "shocker," etc. If cattle go down in a trailer or stocks and refuse to get up, shut off their air by putting

Fig. 10.9—A curved loading chute. Cattle tend to move forward more readily if either a curve or angle is built into the chute so animals cannot look straight through.

a hand over their nostrils. When they attempt to breathe, they will rise.

g. Do not crowd cattle if they are in a corner, especially with wire fencing.

h. Use a sturdy, solid loading chute and handling pens.
i. Make certain nails and other sharp objects are not protruding.
j. Avoid over-crowding and over-heating, especially on hot days.
k. Utilize modern stocks, marking tables, and headgates in performing routine operations on cattle.
l. Provide shade, water, and rest.
m. Let cattle drift from pasture to pasture whenever possible rather than driving them.
n. Handle more than one animal. Cattle become nervous and frightened when alone.

Precondition Feeder Cattle—This means preparing them for the stress of shipping and moving to feed lots. Preconditioning according to the following directions will greatly reduce loss from death, shrinkage, and wasted feed.

a. Wean 30 days before shipment (after castrating and branding).
b. Have a "trough and bunk" adjustment period of 30 days. Give ample clean water. Medicate only on veterinarian's recommendations. Supply 14 per cent protein (total ration) for the 30-day period.
c. Two weeks after weaning, re-vaccinate calves, if they were vaccinated for blackleg, malignant edema, and leptospirosis before 3 months of age. Simultaneously vaccinate for "red nose" (infectious bovine rhinotracheitis), bovine virus diarrhea, and para-influenza 3 (SF-4). Follow your veterinarian's advice for vaccination procedures.
d. Treat for grubs by spray, pour-on, or feed.
e. Check feces for worm eggs; worm if necessary.
f. Wait 10 to 14 days; then move animals to feedlot:

a. by rail 24 hours, then rest them 12 hours.
b. by truck 36 hours, then rest them 12 hours.

Certify the preceding treatments with veterinarian's signature.

Prevent Shrinkage—Shrinkage is the loss in weight between feed lot or farm and market. It is commonly expressed in pounds per head or in per cent of weight before shipment. Some shrinkage is almost inevitable. If approved practices are used as outlined above and in Section 6 on shipping, shrinkage can be held to a minimum. According to the U.S.D.A., the following shrinkage of cattle was determined:

Grain fed—In transit less than 36 hours—3.62% shrinkage.

Grass fattened—In transit less than 36 hours—3.88% shrinkage.

Most shrinkage occurs in the first few hours of shipping. Thus, if cattle are to be shipped 20 miles, little additional shrinkage will occur if shipped several hundred miles. So take advantage of distant markets.

Fig. 10.10—Use partitions to prevent fighting, trampling, and suffocation. Separate sheep, calves, hogs, and different weights of cattle. In every case, tie the bull.

6. Preventing Shipping Losses

Many death losses occur in shipping which are due to improper handling. This can be avoided by careful attention and management. In addition, a great number of shipping losses caused by bruising and crippling can be reduced by use of approved practices in shipping cattle. The following suggestions are offered:

a. Avoid hard driving and allow ample time for rest before loading.
b. On arrival at the pens, do not allow water but provide non-legume hay and allow rest.
c. Do not overcrowd in trucks or cars. Approximately 12 to 15 square feet per mature animal should be allowed depending on age and size.
d. In cold weather, bed the car well. If weather is severe, line the side walls with heavy paper high enough to prevent winds from striking cattle directly.
e. Cover the bottom of the car, truck, or trailer with an inch or so of sand to prevent slipping. In hot weather, wet down the sand to keep cattle cool.
f. Feed, water, and rest at proper intervals en route. Under the 28-hour law, five hours' rest is the minimum specified time.
g. If possible do not load hogs, cattle and young

NUMBER OF ANIMALS FOR SAFE LOADING*

Cattle Per Railroad Car

Average weight	300	400	500	600	700	800	900	1000	1100	1200	1300	1400
36-ft. car	60	50	42	37	33	30	27	25	23	22	21	19
40-ft. car	67	56	46	40	37	33	30	27	25	23	22	21

Cattle Per Truck

Average weight	700	800	900	1000	1100
13-ft. truck	11	10	9	8	7
16-ft. truck	14	13	12	11	10
20-ft. truck	18	16	15	13	12
24-ft. truck	22	20	18	16	15

*Recommended by Livestock Conservation, Inc., Chicago, Ill.

stock in the same truck or car. If they must be transported together a strong partition should separate calves from older stock. Load a short time before leaving.

h. Strong, sturdy loading chutes to load and unload will prevent much injury. Back the truck or trailer squarely against the platform. Do not make cattle "jump" a gap.
i. Truckers should drive carefully, especially starting, stopping, and on turns. Occasional inspection of the load is a desirable practice.
j. Upon arrival, let cattle fill on dry hay; a few hours later supply limited water. By end of first day, give free access to dry hay and water.
k. Consult a veterinarian regarding the use of biological products to prevent shipping fever.

Be Careful of Horns—Special precautions must be exercised with horned animals. In such cases provide more square feet per animal, and valuable purebred stock should be haltered securely and tied. Commercial cattle going to slaughter arrive in better shape if the horns are tipped.

Tie Bulls—Whenever possible it is a desirable practice to tie bulls, especially mature bulls, in the truck or railroad car when transporting them. A stout halter is best although a chain around their necks is satisfactory. Never tie by the ring in the nose. If rope halters are used they should be fitted fairly tight and knotted at the chin strap so the halter cannot slip off.

7. Preparing Cattle for Sale

The fitting of beef cattle for private sale or public auction does not differ widely from fitting of beef cattle for show. However, there are a number of approved practices that are desirable.

Fig. 10.11—Prospective buyers inspecting cattle prior to a sale. Note how cattle are grouped by ages on clean pastures with easy access from one pen to another by use of electric fences and stiles. Easy identification of animals creates an atmosphere conducive to purchase.

Keep Accurate Records—Be specific on all points. Buyers wish to know breeding and calving dates. Guarantees should not be vague and breeders must be willing to back up what they say about their cattle. Herd bulls offered should be presented with a four generation pedigree. For most animals, however, three generation pedigrees are sufficient. Honesty is the best policy, as the reputation of the breeder and his cattle are at stake.

Have Cattle Fitted and Halter Broken—There is no substitute for condition in beef cattle. Fat is appealing and attracts attention. By use of approved practices mentioned in the chapter on feeding, animals can be fed to reach their "bloom peak" at the time of sale. Well-mannered animals always show up better and attract attention from the buyers.

Offer Only Healthy Cattle—Make the Bang's disease and T.B. tests within two to three weeks of the sale. Parasites, such as grubs or lice, should be controlled.

Advertise Well—The amount of advertising will depend somewhat upon the number of cattle to be sold; however, do not neglect this point. Livestock journals, market papers, local and weekly newspapers, and radio are possibilities. Do not neglect neighbors, as they often turn out to be the best customers. If the sale demands a catalog, 500 to 600 copies should be enough. Get the best auctioneer possible.

Have Proper Number of Cattle—As a general rule no sale should contain less than 35 or 40 head. More buyers will be attracted if some variety is offered in age and if both heifers and bulls are available.

Fig. 10.12—A good sale provides for comfort of visitors. These sanitary restrooms solve the problem nicely and can be rented or purchased for the event. They should be placed strategically out of the way and yet be easily visible.

Provide the Proper Setting—The comfort of visitors and prospective purchasers should be considered at all times. A large pavilion or tent is recommended. Lunch and parking facilities should be readily available. Chutes and alleys must be so arranged that cattle can be moved in and out rapidly. A five-foot sale ring made of strong posts and wire netting is desirable in order that everyone can see. Keep the area wet down if dust becomes a problem. The use of extra help during the sale is a desirable practice.

8. Fitting Cattle for Exhibition

Showing cattle at fairs and expositions is one of the best methods of advertising your stock. In addition, the breeder has an opportunity to see how his cattle compare with other cattle of the same breed. Shows

Fig. 10.13—Cattle should be secured with a loose chain around the neck, as illustrated, while being groomed for exhibition.

are also the climax for feeding and fitting fat cattle. 4-H and F.F.A. members can utilize livestock expositions to gain invaluable experiences in feeding and training cattle for show.

Assemble Proper Equipment—Care should be taken to include all equipment needed for showing. A suitable hose and wash rack are necessary. Other equipment includes rope halters, show halters, scotch comb, washing brushes, lining combs, woolen rub rags, grooming oil, soap, water, and feed buckets and forks. The following procedures are recommended in training, grooming, and showing steers.

Teach to Lead—Time, patience, and practice are essential and will be rewarded. A calf should be fed until gentle and then taught to wear a halter. Do not attempt to teach him to lead until he no longer objects to being haltered. Try tying him while he is brushed. One should walk closely to the left side and let the calf move first if necessary. Be firm, but do not abuse or rush him.

Train to Stand Properly—Proper stance is for his feet to be placed squarely under him and well apart. A light stick about 4 feet long with a blunt nail driven cross ways through the end is necessary for this part of the training. The point of the stick may be pressed high between the toes if it is wished to have him move his foot back. The nail may be hooked back of a dew claw and a slight pull will cause him to step forward. He should be taught to keep his head up in an alert position.

Show Animal to Best Advantage—For example, a low-backed steer can sometimes be shown to better advantage if a show stick with a nail in the end is used to make him raise his back. This is accomplished by touching him lightly with the nail under the belly so he tends to raise his back. Try to keep him standing squarely on all feet. He should not get in the habit of completely relaxing while he is being shown, and should be taught to stand at attention for at least 10 minutes at a time. However, an animal should not be worn out by too many strangers handling him or excessive prodding.

Trim the Feet—A steer that carries himself perfectly and stands squarely on all four feet is shown to advantage. Hoofs should be properly trimmed utilizing a pair of hoof nippers and chisel. A good set of stocks is necessary to hold the animal; otherwise it is a most difficult job.

Most breeders trim the bottom of the foot level and as a result the toes are shortened. In no case should the foot trimming job be performed by just shortening the toes and leaving the floor of the foot untouched.

Feet should be trimmed several weeks before show or sale day so that the animal will have sufficient time to recover from any tenderness caused by the trimming. Lame animals are often severely penalized in the show or sale ring.

Wash the Steer—Tar soap is one of the best cleaning agents for washing cattle. The animal should be wet thoroughly and the soap rubbed over him until a good lather results. Care should be taken to rinse all soap out of his hair and water should not be allowed to get into his ears. The steer should not be washed when he is hot. Washing should be done early enough in the morning so he will dry during the sunny part of the day. Once the coat is in good condition, brushing is all that is necessary until about five weeks before the show. After that he should be washed once each week. A weak chlorine or bluing solution may be used to bleach white spots and switches.

Fig. 10.14—Thoroughly wetting down the animal is the first step in grooming for exhibition.

Curl the Hair—The parallel curl is the simplest type and is recommended for beginners. There are different types of curls used in breeding classes, but it makes no difference in fat steer classes because the hair is brushed up until it takes on a soft, fluffy appearance.

Curling is easier if the hair is first wet down by brushing on a solution made of $\frac{1}{4}$ cup of coal tar dip in 3 gallons of water.

The switch can be braided into six 3-strand braids while the hair is damp, tied together, and combed out before the show. Brushing against the hair on the switch will make it appear fluffy. Styles, localities, and breeders may vary as to how the switch is finished.

Clip the Hair—Common practice is to clip the poll of polled breeds. Steers with horns are shown un-

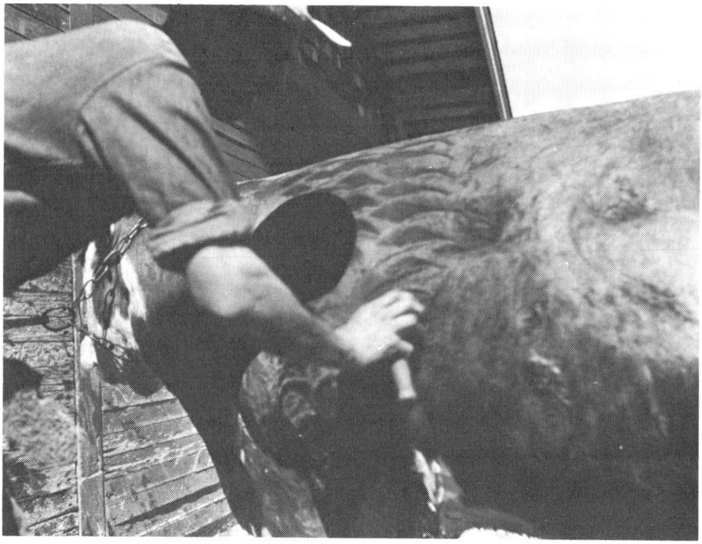

Fig. 10.15—Waving the hair with a zigzag motion is one method of curling hair.

clipped. The tails of all beef breeds should be clipped down to a point where the fullness of the twist begins to give way.

Clip Entire Head of Steers—The entire head of steers for all breeds is clipped to about three inches behind the ear. Do not clip the ear. *Angus breeders* tend to clip the entire head of all animals just as with steers of other breeds.

Polish the Horns—Horns and hoofs can be scraped and sandpapered until they are bright, and a light weight mineral oil rubbed on to brighten their appearance.

Properly Show the Beef Steer—Enter the ring leading the animal from his left side. Hold him close with excess rope coiled. Follow in line moving deliberately.

Fig. 10.16—Clipping the head of an Angus for show purposes. A rope halter can be made quickly and is a handy item to have for handling cattle.

Keep one eye on the judge. Face up hill if possible, but be alert for judge's directions. When standing, square up the legs and straighten the top line by use of show stick on steer's belly. Normally move counter-clockwise, but don't shut off the judge's view. Talk to your animal. Whenever changing positions, back out of line by putting pressure on the animal's shoulder. Be dignified whether you win or lose by continuing to show your animal on the way out.

Consider Your Own Appearance—A good showman is alert at all times. He should dress cleanly and neatly and in keeping with the job of showing an animal.

Fig. 10.17—Georgia students vie for honors with their purebred cattle at one of the many livestock shows. Such events are the culmination of all the effort of fitting.

9. Photographing Beef Cattle

One of the best methods of advertising is by the proper use of good pictures of livestock. Catalogs can be made more effective if they include photographs of animals being offered for sale. In addition, breeders like to know how an animal looked as a calf or how the dam or sire of a calf appeared should it no longer be living. Records are more meaningful if they contain pictures. There are a number of approved practices to be followed in obtaining good photographs.

Properly Fit the Animal—Cattle should be prepared for a photograph just as though they were going to enter the show ring. Proper clipping is important, as mislaid hairs tend to magnify and detract from appearance.

Select the Proper Growth Stage—Animals should be in good condition and healthy. Select a time of year when the hair is long in order to get the best picture. Females look best just before calving. Young, growing animals often appear better one month than the next.

Avoid Distracting Background—Select a site where telephone poles, old machinery, etc., will not be visible. Lawn or a similar covering on the ground is a good location. Landscapes, such as a low, rising hill or hedge, make an ideal setting and do not distract from the animal. Some straw is permissible, but do not stand the animal knee deep in straw, as most people will assume that the breeder is attempting to misrepresent the animal. Feet should be squarely placed and head high.

Pose the Animal—Make certain the animal is alert, by waving a hat or in some way attract its attention.

Beef cattle are generally snapped from the front quarter; this is especially true with bulls. With groups, pictures may be taken from the side or in tandem. Do not take pictures in the wind. Focus directly on the animal. The picture is slightly improved if the background is not sharp. Take the picture a little above the animal; never lower or from a kneeling position. Some animals may appear better on one side than the other.

Keep Records—Age of the animal, stage of lactation, date and location are important if the photograph is to be used for most purposes.

Use Only Good Pictures—Patience will be rewarded. Do not retouch pictures. If the picture does not turn out well, discard it. A poor picture is worse than none.

TYPICAL SHOW CLASSIFICATION* FOR BREEDING BEEF CATTLE

Class	Age of Bulls	Age of Females
Junior calves	Born after Jan. 1 of the current show year. In shows between Jan. 1 and May 1, this class for bulls is divided into spring calves, and those born after April 1.	Same as bulls.
Winter calves	Born between Nov. 1 and Dec. 31 of the previous year.	Same as bulls.
Senior calves	Born between Sept. 1 and Oct. 31 of the previous year.	Same as bulls.
Summer yearlings	Born between May 1 and Aug. 31 of the previous year.	Same as bulls.
Spring yearlings	Born between March 1 and April 30 of the previous year.	Same as bulls.
Junior yearlings	Born between Jan. 1 and Feb. 28 of the previous year.	No class.
Senior yearlings	Born between Sept. 1 and Dec. 31 2 years prior to the show.	No class.
Two-year-olds	Born between March 1 and Aug. 31 2 years prior to the show.	No class.

American Hereford Association

*Show classification will vary according to breed. Associations make available their classification upon request.

Chapter

XI

KEEPING AND ANALYZING RECORDS

The keeping of records and the analyzing of records are most important in beef production, yet are often overlooked. Records should be simple and easily understood by others as well as the breeder. However, they should include all necessary information such as breeding and calving dates, sales and transfers, cost of production, herd improvement records, and the like. There are a number of approved practices in record keeping which should be followed.

Activities Which Involve Approved Practices

1. Keeping records of price trends.
2. Keeping records of breeding dates.
3. Keeping records of calving dates.
4. Using an effective identification system.
5. Keeping weight records.
6. Keeping a proper journal (expenses and receipts).
7. Determining cost of production.
8. Keeping records of sales, transfers, and pedigrees.
9. Analyzing records.

ESTIMATED ANNUAL COSTS PER BEEF COW AND CALF*

	Average Cost	Range of Costs	Your Estimate
Winter feed—roughage	$32.40	$24.00–$44.00	______
Pasture	26.60	18.00– 30.00	______
Grain, protein, salt, mineral	7.90	6.00– 10.00	______
Breeding charge	3.15	2.00– 7.00	______
Vet and medical	3.15	1.50– 4.50	______
Power and fuel	4.20	3.50– 6.00	______
Depreciation and repairs on equipment	.75	.25– 2.00	______
Taxes and insurance	3.00	1.50– 5.00	______
Miscellaneous	1.05	1.00– 3.00	______
Interest	20.50	15.00– 30.00	______
Housing	2.00	0– 5.00	______
Total costs (except labor)	$104.70		

*Includes costs for normal herd replacements. Calf sold at weaning.

Fig. 11.0—The above figures are for Iowa conditions but give some idea of the items and relative importance of each in analyzing a beef cow and calf operation.

1. Keeping Records of Price Trends

Keep Record of Local Current Information—A record of current local prices should be made for each day throughout the year if a careful study of price trends is to be made. This can easily be done by using a simple form and filling in desired information as obtained from the radio, market news, newspapers, and other sources. It is a reasonably easy matter to obtain from U.S.D.A. or agricultural colleges published records of the overall market records for various central markets throughout the nation. It is a desirable practice to correlate local markets with large central markets in order to determine the best time to sell and the best locality to ship. In this manner, prices on one market may indicate prices to be expected on other markets.

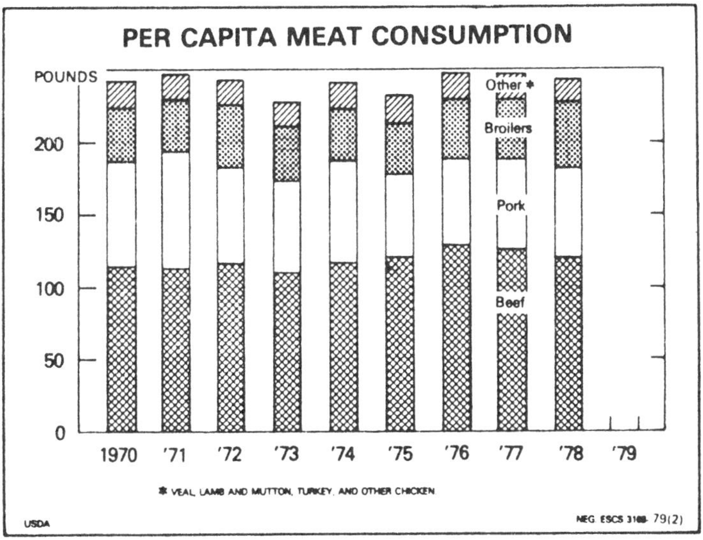

Fig. 11.1—Even though the price of beef has risen steadily, the consumption of beef remains high. Although beef consumption will decline, it will still represent more than 60 per cent of the red meat and over 45 per cent of the red meat and poultry consumed during 1979. Beef supplies will change in composition as a larger proportion of beef will come from fed cattle and much less will come from nonfed cattle. In addition, foreign countries are eating more beef, so the future of the beef industry looks bright.

2. Keeping Records of Breeding Dates

The keeping of breeding dates is important in order that the herdsman knows when the cow is to calve and which bull was used for service.

Use Two Books—Most herdsmen prefer to utilize a small pocket herdbook in which to make entries and records as they occur throughout the day. Periodically this information is transferred to a larger more complete permanent Herd Record Book.

By using the knowledge in his pocket herdbook, the producer can readily check to see when a cow should

BREEDING AND REPRODUCTION RECORD

Cow No.	Name	Date of Calving	Breeding Dates and Bull Used	Date Due	Date of Calving	Herd No. and Sex of Calf	Remarks

come in heat again should fertilization not occur, and later to separate her from the herd as calving time approaches.

3. Keeping Records of Calving Dates

In commercial herds exact records of calving dates for individual animals are seldom made. However pure-bred raisers must know correct dates on which individual calves are dropped. Even in commercial cattle operations exact calving dates are recorded if weight for age records are desired as a guide in developing cattle that are efficient users of feed.

Get Complete Information—All facts concerning the calf should be written down. Sex, number or identification of dam, twin or not, unusual color markings, health, whether calf has been dehorned with paste and so on.

Get Information Early—It is a desirable practice to record information early. Oftentimes death, moving cattle from pasture to pasture, or other causes will prevent accurate records on one individual calf which in turn might confuse the entire herdbook.

Herdsman's Field Sheet

Breeder____

Calf No.	Sex	Birth Date	Dam	Sire	Birth Weight	Weaning Weight	Remarks Weaning Date, Grade, etc.

Fig. 11.2—Calving records are important, especially in performance testing. Herdsmen should carry a pocket field book or have ready access to a field sheet like the one above in order to keep up-to-date accurate records.

4. Using an Effective Identification System

It is extremely desirable that all animals in the herd bear some mark of permanent identification.

Brand Commercial Cattle—All range cattle should be branded. Many purebred livestock are also branded, especially those grown under range conditions. It is most desirable to brand animals as calves before 6 months of age. The most widely accepted method is by the use of a hot iron, although freeze branding is used by some ranchers. Care should be taken to brand deep enough that the scar is permanent but not so deep as to cause a wound slow to heal which will slur the marks so they are not easily read. Brands should only be large enough to read 30 or 40 feet away and located on the less valuable parts of the hide. Unless a brand is recorded with the State Department of Agriculture it has no legal status. See Chapter IV.

Tattoo Calves—The best identity mark is by use of indelible ink tattooed in the ear. The ink should be rubbed well into the tattoo punctures in order to get a

Fig. 11.3—Cattle that have horn brands can be quickly identified without the trouble of catching them.

lasting identification. Oftentimes the herd number and a key to their age are used for a tattoo. It is desirable to tattoo cattle at the same time branding or vaccination is being done.

Horn Brand Mature Cows—With horned breeds it is desirable to burn a brand on one or both of the horns. This is a quick, easy method of identification. Most breeders make horn brands correspond to the animal's ear tattoo. Horn brands cannot be used on yearlings or calves.

Number Individuals—No two animals in the herd should have the same number. Should it not be possible to brand, tattoo, or horn brand, other numbering systems may be used such as ear tags, ear notching, neck

Fig. 11.4—Each animal should be numbered so it can be seen easily and identified. Registered animals are generally tattooed in the ear but they must be restrained in order to be read. Large plastic numbered tags such as the one illustrated are simple to insert, easy to read, and surprisingly permanent. If one is lost or torn out, it should be replaced immediately.

bands and so on; however these systems are not as desirable nor permanent as the former.

5. Keeping Weight Records

Weight records serve as a yardstick for measuring progress in beef cattle production.

Weigh Calves at Weaning—It is a desirable practice to record the weaning weights of calves. This information is an indication of the manner in which the calf can utilize feed, but is especially significant as an aid in determining the excellence of his dam.

Weigh Cattle Regularly—Cattle in the feed lot or range should be weighed every month or two in order to tell accurately how they are gaining. It is necessary to weigh only a relatively small number of animals in order to obtain an indication of the entire herd.

6. Keeping a Proper Journal (Expenses and Receipts)

It is imperative that a complete record of all income and expenditures be kept. This should include a record of all premiums, items sold, or used at home, plus all expenses, repairs, money for labor and so on.

Following is a suggested form in which to keep a record of Receipts and Expenses.*

7. Determining Cost of Production

Each year the records should be carefully reviewed and a cost of production determined for that year.

*Many commercial publishing companies have excellent forms for keeping a record of farming operations. Farm record books are also available from university extension services.

JOURNAL

Date	Description of items, etc.	Cash Book Receipts	Cash Book Expense	Enterprise____ Receipts	Enterprise____ Expense	Enterprise____ Receipts	Enterprise____ Expense
	Total						

Enter amounts twice—once in cash book and also under proper enterprise. This system permits enterprises to be compared with each other.

Cost of production can readily be determined by the following:

$$\frac{\text{Total expense}}{\text{Total yield (head or lbs.)}} \quad \text{Cost per unit}$$

Total expenses should include all costs, such as taxes, interest, and so on in addition to regular items like feed or pasture rent. It is desirable to know the cost of production in order that progress can be measured from year to year.

8. Keeping Records of Sales, Transfers, and Pedigrees

A regular system should be maintained in order to prevent mistakes, unusual delays, and to maintain accurate records.

Register Calves—All desirable calves should be regis-

NAME	NO.	TATTOO
DROPPED	BRED BY	ADDRESS
PURCHASED FROM		COST

PRODUCE RECORD

DATE OF BIRTH			SEX	TATTOO	NAME	REGISTER NO.	SIRE	NO.	TO WHOM SOLD	ADDRESS	PRICE	DATE
MO.	DAY	YR										

SERVICE RECORD

BRED TO	DATE	BRED TO	DATE

GESTATION TABLE—Cows

FIND DATE OF SERVICE IN UPPER LINE
FIGURE BELOW INDICATES DATE DUE TO CALVE

January	1	2	3	4	5	6	7	8	9	10	11	12	13	14	15	16	17	18	19	20	21	22	23	24	25	26	27	28	29	30	31	
October	12	13	14	15	16	17	18	19	20	21	22	23	24	25	26	27	28	29	30	31	1	2	3	4	5	6	7	8	9	10	11	*November*
February	1	2	3	4	5	6	7	8	9	10	11	12	13	14	15	16	17	18	19	20	21	22	23	24	25	26	27	28				
November	12	13	14	15	16	17	18	19	20	21	22	23	24	25	26	27	28	29	30	1	2	3	4	5	6	7	8	9				*December*
March	1	2	3	4	5	6	7	8	9	10	11	12	13	14	15	16	17	18	19	20	21	22	23	24	25	26	27	28	29	30	31	
December	10	11	12	13	14	15	16	17	18	19	20	21	22	23	24	25	26	27	28	29	30	31	1	2	3	4	5	6	7	8	9	*January*
April	1	2	3	4	5	6	7	8	9	10	11	12	13	14	15	16	17	18	19	20	21	22	23	24	25	26	27	28	29	30		
January	10	11	12	13	14	15	16	17	18	19	20	21	22	23	24	25	26	27	28	29	30	31	1	2	3	4	5	6	7	8		*February*
May	1	2	3	4	5	6	7	8	9	10	11	12	13	14	15	16	17	18	19	20	21	22	23	24	25	26	27	28	29	30	31	
February	9	10	11	12	13	14	15	16	17	18	19	20	21	22	23	24	25	26	27	28	1	2	3	4	5	6	7	8	9	10	11	*March*
June	1	2	3	4	5	6	7	8	9	10	11	12	13	14	15	16	17	18	19	20	21	22	23	24	25	26	27	28	29	30		
March	12	13	14	15	16	17	18	19	20	21	22	23	24	25	26	27	28	29	30	31	1	2	3	4	5	6	7	8	9	10		*April*
July	1	2	3	4	5	6	7	8	9	10	11	12	13	14	15	16	17	18	19	20	21	22	23	24	25	26	27	28	29	30	31	
April	11	12	13	14	15	16	17	18	19	20	21	22	23	24	25	26	27	28	29	30	1	2	3	4	5	6	7	8	9	10	11	*May*
August	1	2	3	4	5	6	7	8	9	10	11	12	13	14	15	16	17	18	19	20	21	22	23	24	25	26	27	28	29	30	31	
May	12	13	14	15	16	17	18	19	20	21	22	23	24	25	26	27	28	29	30	31	1	2	3	4	5	6	7	8	9	10	11	*June*
September	1	2	3	4	5	6	7	8	9	10	11	12	13	14	15	16	17	18	19	20	21	22	23	24	25	26	27	28	29	30		
June	12	13	14	15	16	17	18	19	20	21	22	23	24	25	26	27	28	29	30	1	2	3	4	5	6	7	8	9	10	11		*July*
October	1	2	3	4	5	6	7	8	9	10	11	12	13	14	15	16	17	18	19	20	21	22	23	24	25	26	27	28	29	30	31	
July	12	13	14	15	16	17	18	19	20	21	22	23	24	25	26	27	28	29	30	31	1	2	3	4	5	6	7	8	9	10	11	*August*
November	1	2	3	4	5	6	7	8	9	10	11	12	13	14	15	16	17	18	19	20	21	22	23	24	25	26	27	28	29	30		
August	12	13	14	15	16	17	18	19	20	21	22	23	24	25	26	27	28	29	30	31	1	2	3	4	5	6	7	8	9	10		*September*
December	1	2	3	4	5	6	7	8	9	10	11	12	13	14	15	16	17	18	19	20	21	22	23	24	25	26	27	28	29	30	31	
September	11	12	13	14	15	16	17	18	19	20	21	22	23	24	25	26	27	28	29	30	1	2	3	4	5	6	7	8	9	10	11	*October*

Fig. 11.5

SHOW RECORD

Date	Number or Name animal shown	Place shown	Placing	Premium

tered before they are 6 months old. Delay may mean penalties and misunderstandings.

Provide Bill of Sale—It is desirable practice to furnish a bill of sale for all cattle sold and to insist on receiving one when purchasing livestock.

Transfer Promptly—Application for transfer should be made as soon as the cattle are delivered to new owners. Breed associations can supply exact forms and procedures for each particular breed.

Maintain Filing System—It is a desirable practice to maintain an easy to work, simple filing system in order to simplify and speed up transfers, registrations, pedigree information and the like. All records and files should be maintained in safe fire-proof surroundings.

9. Analyzing Records

At the end of the year a careful analysis should be made of the entire cattle operation to see the results accomplished. If records are not analyzed there is practically no reason for going to the trouble of keeping them. Some of the results to summarize should be:

BILL OF SALE
CATTLE, HORSES, SHEEP, (HIDES)

______________________19____

In consideration of________________________dollars paid to me,

by________________________of________________________(State)

I hereby bargain and sell unto above named party________head of________(hides) of which I am the owner.

Animals, (hides) are now at________________________(Address) and described as follows:

No.	Horses Burros Mules	Steers Cows Calves	Sheep	Brand	Location of Brand	MARKS Right	MARKS Left
						( )	( )
						( )	( )
						( )	( )
						( )	( )
						( )	( )
						( )	( )

Seller________________________

Address________________________

Witnesses } ________________________
____

Freeholders of________________________County,

________(State), for two years last past.

Subject to State and City Inspection

Seller agrees to reimburse buyer for losses incurred through inspection.

Fig. 11.6—A typical bill of sale. Livestock producers should keep a supply on hand in order to prevent misunderstandings of livestock transactions. These items are available through commercial printing firms or copied from forms like the above.

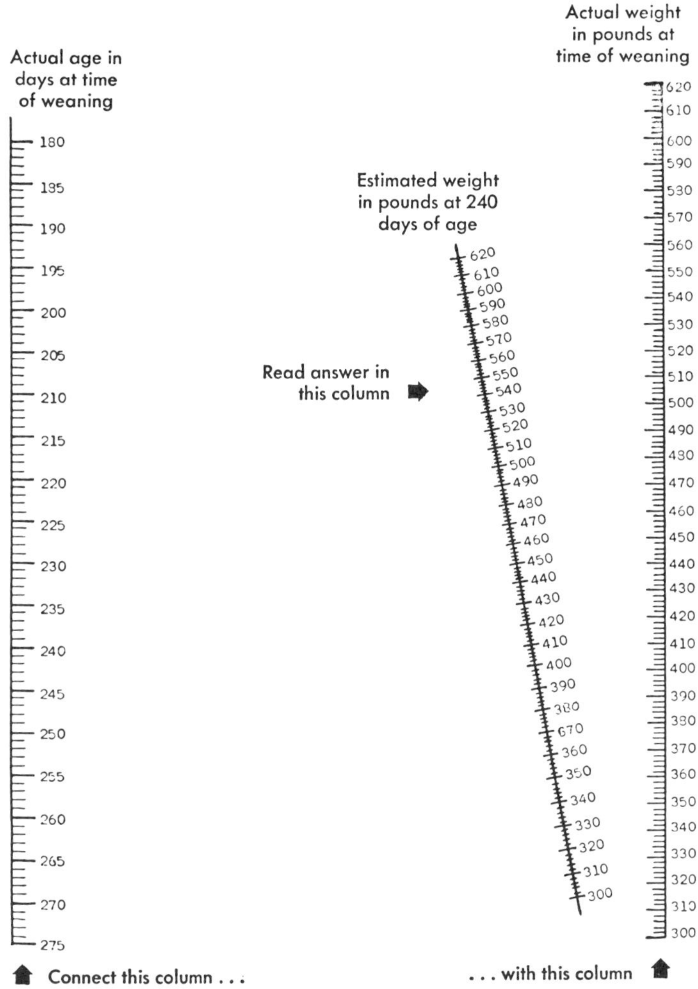

Fig. 11.7—A simple adjusted weight scale. Here's a means of estimating weights of calves at a constant 240 days. Use a straight edge to connect the appropriate figures in the two outside columns, then read the answer in the center column.

a. Determine the percentage calf crop.
b. Determine the cost of producing a pound of beef or veal.
c. Determine the average weight per calf at weaning.
d. Determine the average price received per head for breeding stock, and average price received per pound for fat stock or feeder cattle.
e. Determine average rate of gain for feeder cattle.
f. Determine the number of pounds of beef produced per acre of farm or pasture.
g. Determine the net profit for the cattle enterprise.
h. Determine per cent death loss.
i. Determine the value of rent furnished by producing beef.
j. Determine the amount of labor necessary per unit of beef yielded.
k. Determine the value of beginning and closing inventory.
l. Determine the percentage of miscellaneous expenses such as veterinary, insurance or transportation.

Probably no producer should figure all of these items, but will determine as many as have a direct bearing on his particular spread.

Chapter

XII

PRODUCING BEEF FOR THE HOME LOCKER

Almost $1\frac{1}{2}$ million beef cattle and calves are butchered each year on farms for home use. In addition, more and more people are living on small suburban farms from which they obtain part of their income and a large amount of farm produce for their own consumption. These people work in cities and industrial areas for the greater portion of their cash income. Since economical refrigeration is now available to all, many persons can profitably produce beef for home consumption. The use of approved practices will assist in getting satisfactory results.

Activities Which Involve Approved Practices

1. Determining amount of beef needed.
2. Determining supply of feed.
3. Securing adequate refrigeration.
4. Utilizing nurse cows.
5. Utilizing waste feeds.
6. Purchasing the right feeders.

1. Determining Amount of Beef Needed

The average consumption of beef per capita in the

United States is about 70 pounds plus 10 pounds of veal. Therefore, a family of five would require 350 pounds of beef and 50 pounds of veal, or about 400 pounds per year. Persons doing much physical work and rural people in particular would require considerable more meat for their needs.

One Steer Per Year Needed—The average dressing percentage of beef cattle is roughly 52 to 55 per cent so that one 800 or 900 pound steer would provide enough beef for a family of five. A desirable practice is to produce slightly more beef than the home can use. The extra meat can be ground or sold so that only the more desirable cuts will be available to the farm family.

Use Other Meats—Lamb, pork, fowl, and other meats should also be produced or available so the family will not tire of too much beef and thus better utilize the various cuts of beef or veal.

2. Determining Supply of Feed

Determine Feed First—It may not be profitable for all small farms to produce their own beef. Therefore, it is essential to first determine whether an adequate supply of economical feed is available. Many people commit the error of getting the cattle first.

Determine Amount of Feed—The average cow will require about 3 to 3½ tons of good hay and 1 ton of concentrates per year. If this amount cannot be purchased economically or raised cheaply, other sources of feed should be investigated.

Utilize Irrigated and Permanent Pastures—If land is available, this source of nutrients is perhaps the most economical and satisfactory way to supply feed.

Generally speaking, $1\frac{1}{2}$ to 2 acres of growing pasture is adequate to keep one animal and its offspring. Additional feed may have to be purchased depending on locality.

It is a desirable practice, especially with irrigated pasture, to divide the pasture into two or more fields so it can be properly managed. Oftentimes a single-strand electric fence is suitable if there is a good outside fence.

Cut Excess for Hay—If the field is slightly larger than that required for grazing, the extra feed can be cut for hay during the rapid growing season. This will also prevent the pasture from becoming rank and of poor quality in addition to reducing weeds.

3. Securing Adequate Refrigeration

If meat is to be kept for any length of time, it is imperative at present that storage facilities be of the freezer style that keeps meat continually below freezing.

Obtain Adequate Size—As a general rule it is better to get a freezer of larger capacity than just the calculated size. Larger freezers do not have to over-work in order to keep meats safe and they take care of unexpected emergencies. The cost of a slightly larger size is seldom much more on a percentage basis of the total and is well worth the added cost. It is a desirable practice to consult the manufacturer of freezers in determining proper size for a given family need.

4. Utilizing Nurse Cows

In order to obtain excellent quality beef with a minimum of hand labor, the steer or heifer should be allowed to nurse its dam as long as possible. Extra feed or good grazing should be provided the dam in order to maintain her body and supply enough milk for the calf.

Raise More Than One Animal—Most cattle are nervous and do not do as well alone as compared to having another animal around to quiet them. This is not as noticeable with calves but is sometimes an important factor with older feeders that are fed out singly near to the farmstead. If the feeders have been accustomed to running with a large herd consideration should be given to this item. It is a desirable practice to feed out more than one animal. A dairy-type cow that can be used for milk purposes and a nurse cow as well will satisfy the feeder's desire for companionship and prevent it from becoming too nervous. It is often a good practice to fatten two feeders and then market one to help defray costs.

Use Dairy-Type Cows—If beef cows are not available, dairy cows can raise excellent beef calves and if bred to beef bulls produce reasonably good crossbred calves to raise for beef or veal.

Crossbred calves of such matings will not yield carcasses equal in quality to pure beef breeds; however they are adequate for the average table, especially in localities where taste does not demand fat beef.

This is an especially desirable practice as one good dairy-type cow can supply sufficient fluid milk and its products for a family and at the same time nurse about one beef per year providing both the cow and her calf have access to good grazing. If the cow is selected for the right temperament, she will allow any calf to nurse any time and thus eliminate the worry of finding a milker should the family go away on vacation or other duties prevent regular milking.

5. Utilizing Waste Feeds

On many farms and on part-time farms in suburban communities, there are waste feeds available that can profitably be fed to cattle. Lawn clippings, waste materials from stores and markets or processing plants, as well as excess vegetable or crop products from the farm can be turned into meat.

Feed Only Unspoiled Products—The greatest danger in utilizing waste feeds is the presence of poisoning due to spoilage. Even if cattle are actually not poisoned, they may go off feed. Therefore, it is a desirable practice to feed only fresh waste products or crops that have been properly cured.

6. Purchasing the Right Feeders

When only one or two animals are being purchased, it is sometimes difficult to find an adequate source of feeders. Often neighbors are the best source because of off-season calves or orphans. Public auctions, central livestock yards, and private sales generally can supply feeders to those who wish to purchase only one or two head. Calves or feeders of pure beef breeding are most desirable but oftentimes are not readily procured in suburban areas, and for this reason crossbred calves with some dairy-type blood are the only animals available. However, as previously stated, these animals will make satisfactory beef for average home use if ample feed is available.

Purchase Disease Free Stock—Particular attention should be given this item because an unhealthy condition may be the reason such animals are offered for resale. Weak or sickly calves or feeders may never recover or, if they do, make gains too expensive even though initial cost be low.

Do Not Purchase Fat Cattle—Cattle or calves that are in extra-good condition are not desirable to feed out under home-use conditions. Generally, these cattle have reached their peak and should be slaughtered. Such animals will often go down hill when taken home to raise due to the change in ration, handling, weaning from nurse cow, and other reasons that make ultimate gains too expensive.

Select Animals That Are Growing—Generally speaking, this means animals that are young and healthy but in only fair condition. This type will not be affected so much by changes in their ration and are still gaining in weight due to body growth. If rapid gains are desired, older animals will consume more and make a large daily gain. However, the older an animal gets, the more expense it is to put on weight. On the other hand, many people prefer the taste of mature cattle and would therefore do well to purchase older, thinner feeders to fatten to their particular taste.

Chapter

XIII

ESSENTIAL SKILLS FOR CATTLEMEN

In order to successfully carry on a beef producing enterprise, the operator should have a large percentage of the following skills and abilities. Skills are generally considered the manipulative things that a producer must be able to do with his hands; abilities are most often considered to be activities involving thinking only.

The beef producer should be able to:

(1) **General.**

a. Select feeder calves.
b. Select yearling bulls and heifers.
c. Plan a feeding schedule for steers.
d. Plan a feeding schedule for growing heifers and breeding cows.
e. Determine economical feeds and compile ration.
f. Tattoo a calf.
g. Throw or "cast" an animal in order to perform livestock operations or treat for disease.
h. Castrate calves.
i. Dehorn calves or mature animals using caustic, saw, or clippers.
j. Make a rope halter.

k. Recognize a cow in heat.
l. Remove foxtail from eye or wound.
m. Assist cow at calving.
n. Treat for bloat.
o. Determine what to do if cattle are off feed and scouring.
p. Treat for internal parasites.
q. Treat for cattle grub and other external parasites.
r. Treat for removal of warts.
s. Treat for foot rot.
t. Vaccinate to prevent disease.
u. Prevent shipping fever.
v. Implant feeder cattle.
w. Dip or spray cattle.
x. Build fence.

Fig. 13.0—Illustrating method of "casting" a large animal. The rope is thrown around the animal's body several times, as indicated. When pressure is exerted by pulling on the rope towards the rear, it will cause the animal to lie down.

y. Plan and construct loading chute and restraining equipment.
z. Bridle, saddle, and ride a saddle horse.
aa. Determine the age of cattle by their teeth.

(2) **Additional skills desirable for purebred producers, but helpful to all beef producers.**

a. Clip tail and head.
b. Polish and train horns.
c. Curl and wash for show.
d. Train animals for show.
e. Trim feet.
f. Ear tag.
g. Ring a bull.
h. Grind and mix feed.
i. Register an animal and make transfer papers.
j. Fill out entry blanks for stock show.

A new breeder would not be called upon to do all of the skills listed. Proficiency will come with practice and experience. However, the more able the operator can become in performing needed operations, the more efficient his operations will be with a consequent lowering in cost of production.

(3) **Determining the age of cattle by their teeth.**

As the productive life of farm animals is comparatively brief, the age of an animal is therefore of considerable importance to the livestock producer. One of the most accurate ways of determining age of cattle is by examination of the teeth. These methods are shown in the following illustrations.

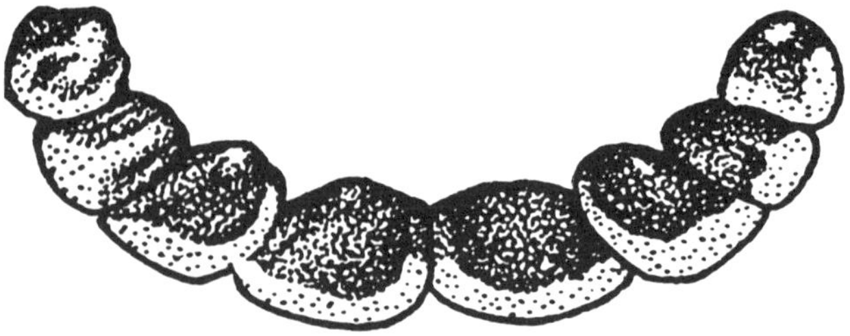

Fig. 13-1—Internal face of incisors of the calf at 1 month.

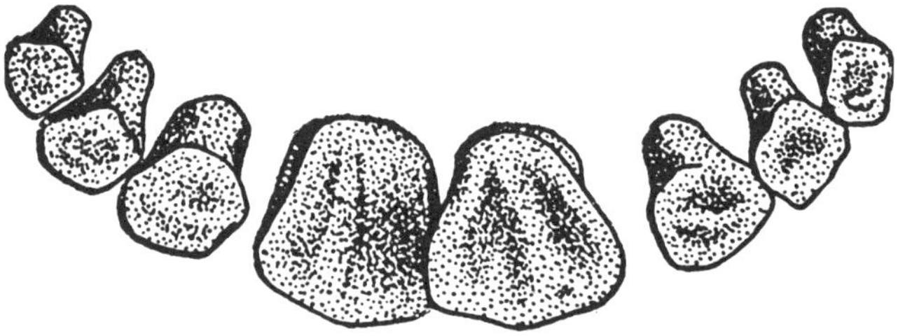

Fig. 13.2—Internal face of incisors at 2 years.

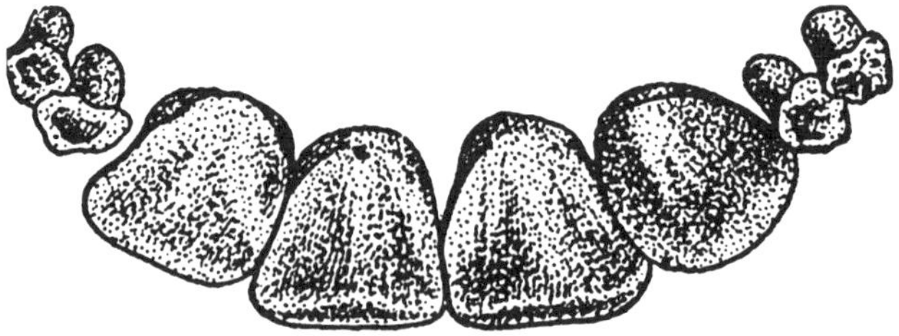

Fig. 13.3—Internal face of incisors at 3 years.

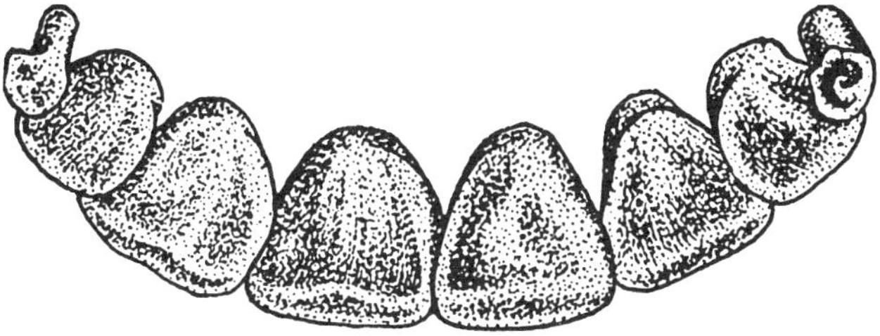

Fig. 13.4—Internal face of incisors at 4 years.

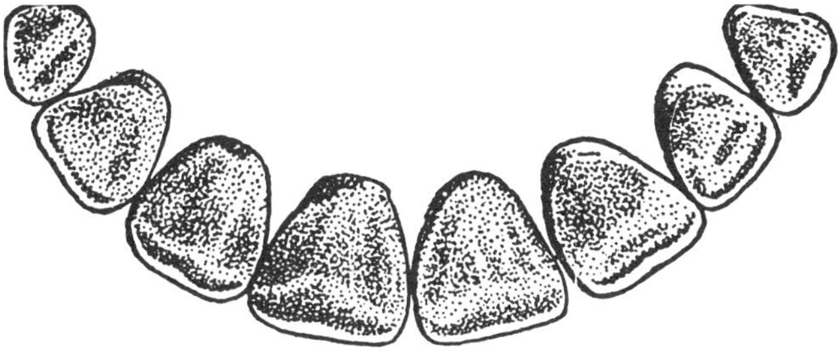

Fig. 13.5—Internal face of incisors at 5 years.

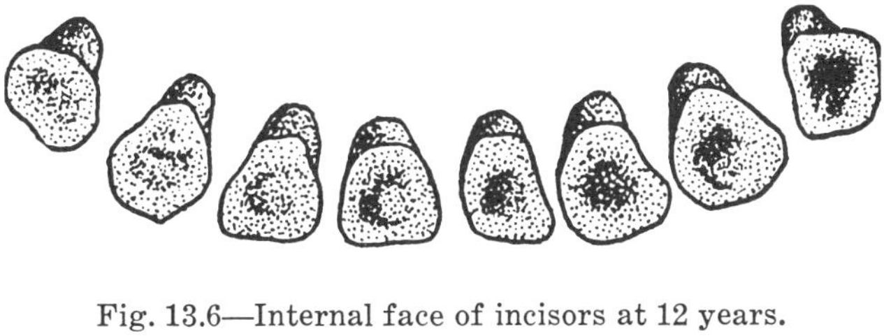

Fig. 13.6—Internal face of incisors at 12 years.

(4) Determining the Age of Calves.*

Fig. 13.7—Senior calf. All eight temporary incisor teeth firmly in place. One week less than 15 months of age.

Fig. 13.8—Summer yearling. All temporary incisors still firmly in place but "necking" out close to the gums. No evidence of permanent teeth. One week less than 19 months of age.

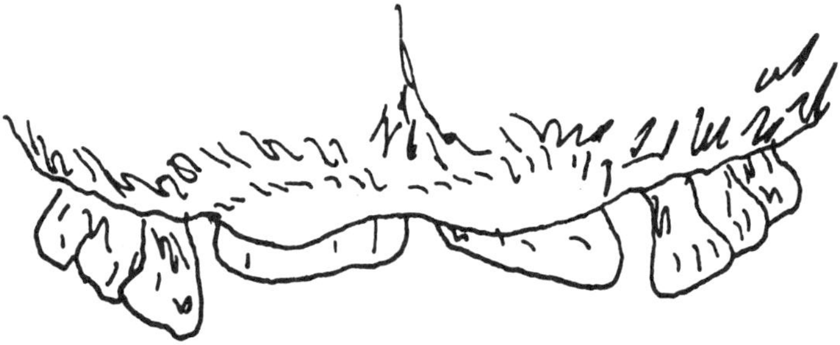

Fig. 13.9—Junior yearling. Two temporary incisors have just been replaced by permanent incisors. New permanent incisors need not be in wear. One week less than 23 months of age.

---

* Iowa Farm Science, Vol. 19, No. 1

Fig. 13.10—A desirable way to throw a calf, one of the skills every stockman should learn. Note especially how the calf is gripped prior to actual throwing. Throwing is accomplished by lifting with the hands and shoving in with the knees.

APPENDIX A

Beef Cattle Associations

Angus

American Angus Association
3201 Frederick Boulevard, St. Joseph, Missouri 64506

Beefmaster

Beefmaster Breeders' Universal
Suite 720 GPM South Tower, 800 NW Loop 410
San Antonio, Texas 78216

Belted Galloway

Belted Galloway Society
P. O. Box 5, Summit, Ohio 43962

Brahman

American Brahman Breeders' Association
1313 La Concha Lane, Houston, Texas 77054

Brangus

International Brangus Breeders' Association
9500 Tioga Drive, San Antonio, Texas 78230

Charbray

American Charbray Breeders' Association
1610 Old Spanish Trail, Houston, Texas 77054

Charolaise

American International Charolaise Association
1610 Old Spanish Trail, Houston, Texas 77054

Chianina

American Chianina Association
P. O. Box 159, Blue Springs, Missouri 64015

Devon

Devon Cattle Association
P. O. Drawer 628, Uvalde, Texas 78801

Galloway

Galloway Cattle Society of America, Inc.
Hennepin, Illinois 61327

Gelbvieh

American Gelbvieh Association
202 Livestock Exchange Building
Denver, Colorado 80216

Hereford

American Hereford Association
715 Hereford Drive, P. O. Box 4059
Kansas City, Missouri 64101

Limousin

North American Limousin Foundation
100 Livestock Exchange Building
Denver, Colorado 80216

Maine-Anjou

American Maine-Anjou Association
564 Livestock Exchange Building
Kansas City, Missouri 64102

Murray Grey

American Murray Grey Association
1222 North 27th Street, Billings, Montana 59102

Polled Hereford

American Polled Hereford Association
4700 East 63rd Street, Kansas City, Missouri 64130

Polled Shorthorn

Polled Shorthorn Society of the American Shorthorn Breeders' Association
8288 Hascall Street, Omaha, Nebraska 68124

Red Angus

Red Angus Association of America
P. O. Box 776, Denton, Texas 76201

Santa Gertrudis

Santa Gertrudis Breeders' International
P. O. Box 1257, Kingsville, Texas 78363

Scotch Highland

American Scotch Highland Breeders' Association
P. O. Box 249, Walsenburg, Colorado 81089

Shorthorn

American Shorthorn Breeders' Association
8288 Hascall Street, Omaha, Nebraska 68124

Simmental

American Simmental Association
Simmental Way, Bozeman, Montana 59715

Texas Longhorn

Texas Longhorn Breeders' Association of America
204 Alamo Plaza (Menger Hotel)
San Antonio, Texas 78205

Dual Purpose Cattle Associations

Milking Shorthorn

America Milking Shorthorn Society
17720 JJ South Glenstone Avenue,
Springfield, Missouri 65802

Red Poll

American Red Poll Association
3275 Holdrege Street, Lincoln, Nebraska 68503

National Farm Associations

American Farm Bureau Federation
225 Touhy Avenue, Park Ridge, Illinois 60068

National Cattleman's Association
1001 Lincoln Avenue, Denver, Colorado 80201

National Council of Farmer Cooperatives—
Allied Grape Growers
1495 West Shaw Avenue, Fresno, California 93705

National Farmers' Union
12025 East 45th Avenue, Denver, Colorado 80251

The National Grange
1616 H Street, N.W., Washington, D.C. 20006

Livestock Magazines Associated with Cattle Industry

Arizona Farmer Ranchman
2214 North Central, Suite 101, Phoenix, Arizona 85004

California Cattleman
P. O. Box 1618, Auburn, California 95603

The Cattleman
Texas and Southwestern Cattle Raisers' Association, Inc.
410 East Weatherford Street, Forth Worth, Texas 76102

Farm Journal
230 West Washington Square
Philadelphia, Pennsylvania 19105

Journal of Animal Science
American Society of Animal Science
309 West Clark Street, Champaign, Illinois 61820

Polled Hereford World
4700 East 63rd Street, Kansas City, Missouri 64130

Western Livestock Journal
Crow Publications, Inc.
4701 Marian Street, 4th Floor, Livestock Exchange Building
Denver, Colorado 80216

Western Livestock Reporter
P. O. Box 2559, Billings, Montana 59107

Organizations Allied with Livestock Production

American Meat Institute
P. O. Box 3556, Washington, D.C. 20007

California Beef Council
1436 Rollins Road, Burlingame, California 94010

Livestock Conservation Institute
1100 Jorie Boulevard, Suite 143
Oak Brook, Illinois 60521

National Association of Animal Breeders
P. O. Box 1033, 401 Bernadette Drive
Columbia, Missouri 65205

National Livestock Exchange
104 South Muskego Avenue, Milwaukee, Wisconsin 53233

National Live Stock and Meat Board
444 North Michigan Avenue, Chicago, Illinois 60611

National Livestock Producers Association
307 Livestock Exchange Building, Denver, Colorado 80216

National Society of Livestock Record Association
210 Utah Avenue, West Plains, Missouri 65775

U.S. Department of Agriculture
Office of Information, Washington, D.C. 20250

APPENDIX B

Answers to Brand Quiz (Fig. 4.15)

1. D Slash D
2. Y Hanging Lazy V
3. Circle Bar
4. Bar Flying V
5. Diamond B
6. Diamond K
7. Half Circle O Connected
8. Ox Yoke T
9. Triangle Diamond
10. Forty Five
11. O Bar O
12. Bar X
13. Running N Bar Running N Bar
14. Broken Arrow
15. L Bar S
16. Box B
17. I U
18. Diamond J
19. Three O Rail
20. P Triangle
21. Flying U
22. Rafter X
23. Bench L
24. Circle B
25. Striped S
26. O I or Cinch Buckle
27. Spearpoint 4
28. X Diamond Bar
29. Lazy T 2
30. N Lazy F Connected
31. R Bar 4
32. Cross C
33. Lazy Y J
34. V Bar Open A
35. K Bar U
36. Y 6 Quarter Circle
37. Bar C Bar
38. J Lazy H
39. Horseshoe Dot
40. Lazy B Swinging L
41. Star M
42. 2 Bar 2 Slash
43. Double Circle (Donut)
44. 81 Bar
45. Crescent Slash
46. Butcher Block
47. Half Circle L
48. 15 Lazy T
49. K 5 Connected
50. T 4
51. Reverse 2 Standing N
52. Lazy S L
53. Lazy E A
54. O W
55. Cross Heart
56. H Swinging Reverse S

APPENDIX C

Summary of Approved Practices in the Beef Enterprise

CHAPTER I

Opportunities in Beef Production

1. An economical supply of feed must first be obtained before entering the beef industry.
2. A commercial herd should be developed first before entering the purebred business.
3. Management factors and other means of lowering costs of production must be practiced rather than increasing size of operations in order to be successful.
4. Waste crops and rough land should be made to produce a salable product through beef cattle.
5. Permanent pastures should be utilized in most areas in order to produce cattle economically.
6. Part-time farmers can economically feed one or two beeves per year in order to supply food for home consumption.
7. Cow-calf operations are possible on a feed lot basis.

CHAPTER II

Selecting the Breeding Stock

1. The type of cattle selected should be early maturing, rugged and capable of weighing 700-1000 lbs. at 12 to 18 months of age under good conditions.
2. A breed should be selected that best meets local conditions. Personal preference should be an important consideration in selecting a breed.
3. Select foundation cows first, then obtain a bull that can improve the offspring.
4. More emphasis should be given to the individuality and performance of a mature animal than to its pedigree.
5. Select a herd sire that will improve the weak points of the cows.
6. Older bulls can be selected on their breeding records alone. A young bull should be selected both from his pedigree and his individuality.
7. Buy proven bred mature cows with a calf when starting a herd.

CHAPTER III

Breeding and Improving Beef Cattle

1. Use high quality purebred bulls to grade up herds. Purebred breeders must use special care to cull rigidly.
2. Determine the size of the herd by the amount of available feed.
3. Select replacement heifers on their capacity to consume feed and their respective weights at one year of age.

4. Heifers should be bred according to their size. Well grown heifers should weigh at least 650 to 700 lbs. when bred.
5. Maintain a short breeding season of around 3 months so the calf crop will be uniform in size. Calves should be dropped as early as weather conditions will permit.
6. In purebred herds one bull may serve as many as 50 cows. On the range use 3 or 4 bulls for every 100 cows.
7. Bulls should be rotated every 10 to 14 days.
8. Bulls should gain weight during the winter so they will be in thrifty condition at the start of the breeding season.
9. Divide females into age groups—the cows in one group and the heifers in another.
10. During weaning and calving time cows must be fed to gain at least the weight lost in producing their calves.
11. Keep records of performance in order to measure improvements in breeding beef cattle.

CHAPTER IV

Production of Calves and Feeder Cattle

1. Separate the pregnant cows from the other cattle. A well-drained pasture or well-bedded box stall should be provided depending on weather in which to calve. Provide ample water.
2. Reduce the amount of feed as parturition approaches.
3. Do not bother the cow when she is about to calve, but be on hand to render assistance should she have difficulty.
4. Make certain the calf is breathing and gets some

of the "first" milk. Disinfect the navel and apply fly repellent to it.

5. Dispose of the afterbirth. If it is retained longer than 24 hours call a veterinarian.
6. Cows should be fed lightly at first. Good green pasture is the best guarantee of health.
7. Heifers bred too young should have their calves vealed so they can resume normal growth.
8. Record the date of birth in the herdbook.
9. All range calves should be branded before they are four months old. Hot iron branding is the most permanent method.
10. Tattoo all purebred calves on the inside of their ears. Dehorn all commercial calves with dehorning paste when they are 2 to 3 days old.
11. Practice creep feeding calves. Well bred calves properly creep fed should weigh 700 to 800 pounds at 10 to 12 months of age.
12. Salt should be provided "free-choice" at all times. Other minerals may be necessary depending on local conditions.
13. Calves should be "gentled" and their horns properly trained if they are to be sold as individual breeding animals.
14. Separate the bull calves from heifers at about 6 months of age.
15. Castrate bull calves before they are weaned. They should be castrated as young as possible.
16. Vaccinate calves against blackleg before they are turned onto the range.
17. Provide good pasture for the growing herd as well as the breeding cattle. Green legume grass type pastures are recommended.
18. Wean calves before winter when they are 6 to 9 months of age.
19. Utilize the principle of continuous growth by

supplementing poor grazing conditions. Calves properly wintered should weigh 500-600 pounds in the spring and 800-900 pounds at the end of the following pasture season.

CHAPTER V

Feeding Beef Cattle

1. Provide sufficient roughages at all times. Permanent pastures are generally the cheapest source of nutrients.
2. Provide a mixture of grasses and legumes in pastures in order to reduce bloat.
3. Purchase high quality hay as the cost per pound of nutrient is often less than in poor quality hay.
4. Silage is an excellent feed for wintering beef. A good rule is to feed 3 pounds of silage for every 100 pounds live weight of animal.
5. Provide concentrates to fatten beef. Corn is the single most important crop used in fattening cattle.
6. Purchase both protein and carbohydrate feeds on the basis of cost per unit of protein or carbohydrate.
7. Minerals must be provided depending on the particular minerals lacking in local areas.
8. Mature cattle will need about 10 gallons of clean water daily. In extreme freezing weather it should be warmed slightly.
9. Grind small grains and corn and cob meal. However, if hogs are following cattle there is no advantage in grinding shelled corn.
10. An ample supply of mother's milk is the best way to raise the beef calf.

11. Creep feed calves if their dams are poor milkers or grazing conditions inadequate.
12. Sort the breeding herd, especially in winter, into various groups according to size, sex, strength, and condition for best feeding practices.
13. Supplement feed nursing cows daily with approximately 2 pounds concentrates in the winter.
14. Generally bulls should be fed a protein supplement, such as cottonseed, about 60 days before the breeding season.
15. In fattening cattle select a ration that will utilize the cheapest feeds available without materially changing the effectiveness of it.
16. Follow cattle with hogs when feeding corn. One 100-pound hog should be provided for every 1000-pound steer.
17. Do not change rations abruptly. Any change in ration or increase in amount should be done gradually over a one or two week period.
18. Provide some coarse roughage in order to reduce bloat when pasturing green growing pastures such as alfalfa or clover.
19. When selecting a calf to fit for show choose fancy or choice calves from a herd known to produce good feeders.
20. Feed more than one calf as single animals become restless and will not "do" as well.
21. Weigh calves regularly every three or four weeks. Commercial feeders need only to weigh a representative number of herd.
22. Provide for economical use of labor by letting cattle harvest their own feed whenever possible.
23. Provide shade for beef animals and eliminate flies with modern fly killers.
24. Protect cattle from poisonous plants by provid-

ing adequate feed and eliminating dangerous areas.

CHAPTER VI

Providing Shelter and Equipment

1. Use 4-strand barbed wire fence on the range; however, only strong board fencing should be used wherever cattle must be "crowded."
2. Treat posts with a preservative in order to prevent rot.
3. Use electric fences to rotate cattle in irrigated pastures.
4. Locate corrals near watering places.
5. Install cutting gates in order to separate cattle easily.
6. Provide permanent type water troughs with automatic water regulators.
7. Provide adequate protection against storms. Cattle can stand cold if they are dry. Sheds should have a southern exposure.
8. Provide windbreaks 8 feet high and at right angles to prevailing winds.
9. Storage capacity for feed should be slightly more than the amount necessary to carry the herd through the winter.
10. Provide self feeders for concentrates and hay racks for feeding hay. Most equipment should be mounted on skids.
11. Loading and restraining equipment such as a squeeze should be used in handling cattle.
12. Locate scales close to corrals and loading chutes. Construct ample runways to facilitate handling.
13. Purebred breeders should provide wash racks

that are located out of the wind and open to the sun.

CHAPTER VII

Controlling Parasites and Diseases

1. Practice sanitation. Eliminate danger from germs by use of proper disinfectant.
2. Provide safe pastures. Dangerous areas should be fenced off. If bogs are present they should be drained.
3. Eliminate mechanical hazards by collecting old wire nails, weakened gates and so on that would permit animals to cut themselves.
4. Keep animals healthy and well fed. Healthy livestock can resist disease more effectively and seldom eat poisonous plants.
5. Destroy all dead carcasses regardless of cause of death.
6. Supply some coarse roughage at all times in order to keep cattle healthy and reduce danger of bloat.
7. Consult a veterinarian immediately on any serious disease such as anthrax or shipping fever.
8. Eliminate flies by treating fly breeding places and use of modern fly killing chemicals.
9. Provide a systematic program of calfhood vaccination against such diseases as blackleg and Bang's disease.
10. At the first sign of cattle being sick, a good measure is to reduce feed—especially concentrates. Separate sick animals.
11. Be alert at all times to discover choke or bloat early if cattle are grazing root crops.
12. Cooperate fully with Federal and local authorities in suppressing spread of disease.

CHAPTER VIII

Butchering Beef on the Farm

1. Select good type yearling farm animals for farm slaughter. 800- to 900-pound animals are desirable.
2. Assemble all equipment such as a hoist, knives, etc. Have plenty of clean water available.
3. Withhold feed for 24 hours before killing. However, supply plenty of water.
4. Securely snub the animal before killing. Kill by stunning or shooting.
5. Stick the beef while it is still on the ground and pump out the blood by alternately pushing a foot into its flank and pulling its tail.
6. Skin the animal carefully so that the hide is not punctured and little or no meat is left clinging to the skin.
7. Tie off the gullet with a stout cord when removing the viscera.
8. Wash and shroud the carcass after splitting it into two halves.
9. Ripen beef by allowing it to hang 7 to 10 days in a temperature of 32 to 36 degrees.
10. Sell the hide immediately or else salt it well if it cannot be disposed of till sometime in the future.

CHAPTER IX

Selecting and Using Beef

1. Select meat from a carcass that has short shanks and neck, well-developed round, heavily fleshed loin and ribs and with not too large a

proportion of chuck and plate.
2. A well finished carcass should have an even covering of brittle, flaky, creamy white fat over the chuck, ribs, and loin.
3. Select a well "marbled" carcass which has relatively large deposits of fat between the large muscles and between the muscle bundles.
4. Select meat from a beef carcass that has relatively small bones.
5. Mature animals, if of proper conformation, will be well flavored and yet tender.

CHAPTER X

Marketing Beef

1. Study the markets carefully before deciding where to sell cattle. Local markets may often pay as much as more distant points.
2. Purebred producers must develop their own market. A good reputation is imperative in this phase of the industry.
3. As a general rule market cattle early in the season. Upgrade and sell replacements.
4. Determine demand before shipping cattle by contacting commission firms or dealers. One or two days may make a considerable difference in prices.
5. When cattle are in peak condition they should be sold unless very unusual conditions prevail.
6. Sort cattle into uniform groups as to quality, weight, age, sex and so on before offering for sale.
7. Carefully handle cattle during loading and transit so they will arrive in top market condition.
8. Have specific breeding and calving dates as well

as other pertinent information available to buyers of purebred cattle.

9. Offer only healthy cattle for sale. Cattle will also make a better showing if halter broken.
10. Advertise a sale well and offer no less than 35 or 40 head at one sale or auction.
11. Make provision for comfort of buyers by having lunch and other facilities available.
12. Teach animals to lead and stand properly when fitting them for exhibition.
13. Wash show steers with a tar soap and curl the hair till it becomes soft and fluffy.
14. The showman should be alert and properly dressed, according to the nature of his job.
15. Photograph only beef animals that are properly fitted and posed. Select the growth stage at which they appear their best. Use only good photographs.

CHAPTER XI

Keeping and Analyzing Records

1. Correlate local market conditions with central markets so as to determine the best time to sell.
2. Keep a small pocket size herdbook for daily entries and a complete larger permanent herdbook as well. Include calving dates, breeding dates, weaning dates and other basic information.
3. Effective identification such as branding, tattooing or ear tagging is imperative in maintaining proper records.
4. Number every individual in the herd. No number should be used twice.
5. Weigh all calves at weaning in order to determine excellence of the dams.

6. Keep a complete record of all expenses and income in a recommended journal. Make entries promptly.
7. Determine cost per unit of production by dividing total expense by total yield.
8. Register all desirable calves before they are 6 months old. Transfer all cattle sold promptly.
9. Analyze records yearly in order to determine future progress.

CHAPTER XII

Producing Beef for the Home Locker

1. Feed out one 800- to 900-pound steer yearly for the average family of five.
2. Determine whether or not feed is available before getting a steer to raise.
3. The average mature animal requires about three and one-half tons of hay and 1 ton of concentrates per year. Younger animals will require less.
4. Provide adequate refrigeration. A slightly larger capacity than the calculated size is a desirable unit to obtain.
5. Use dairy type nurse cows both for raising the beef calf as well as providing fluid milk for the home.
6. Breed the home dairy cow to a beef bull to produce a fairly good calf suitable for slaughtering for home use.
7. When purchasing feeders select young animals that are still growing.

CHAPTER XIII

Essential Skills for Cattlemen

1. Learn to perform essential skills as quickly as possible. See Chapter XIII for complete list.

APPENDIX D

Glossary of Beef Terms

Acorn calves—Calves born deformed as a result of maternal nutritional deficiencies.

ADG—Average Daily Gain (e.g., for 140 days).

Adjusted 205 day weight—A standardized weight to facilitate comparisons, determined as follows:

$$\frac{\text{actual weaning wt.} - \text{birth wt.}^*}{\text{weaning age in days}} \times 205 + \text{birth wt.}$$

*A constant value may be used if birth weight is unknown, e.g., 70.

Anabolic—Of or promoting anabolism; the process in plant or animal by which food is changed into living tissue.

Artificial insemination—The process of mechanically introducing semen into the reproductive tract and uterus of a female for the purpose of producing fertilization.

Baby beeves—Fat cattle 9-15 months old weighing 700 to 900 pounds, usually of high market grade.

Balling gun—Metal instrument inserted in throat to discharge pills into throat.

BCIA—Beef Cattle Improvement Association.

BIF—Beef Improvement Federation.

Fig. D.1—Acorn calf. Born deformed as a result of maternal nutritional deficiencies.

Blocky—Conformation that is deep, wide, and low set. A compact animal.

Bolus—A large pill. Generally a mass of material ground and compressed into a ball, e.g., cud.

Brand—Permanent identification on animals caused by scar tissue on skin, usually performed by hot iron, chemicals, or super cold, as in freeze branding.

Branding—Act of putting on a brand.

Bred—Animal that is pregnant.

Breed—A group of animals possessing certain well-defined distinguishing characteristics and which are able to reproduce these characteristics in their offspring with a reasonable degree of certainty.

Breed character—A combination of masculinity or

femininity with ideal breed type features. Head and color markings are given considerable attention in sizing up breed character.

Breed type—A particular form typical of the breed together with those special characteristics in head and color markings which are common to a particular breed.

Breeding value—An estimate of what an animal is expected to transmit to his or her offspring for a particular trait, such as weaning or yearling weight.

Brockle faced—White-faced cattle with other colors splotched about the head and face.

Bull—Male of cattle; generally used for breeding purposes.

Bulldog calf—Hereditary condition in which calves are born dead, with extensive deformities, and generally born prematurely. (Calif. Bull. 699)

Bulldogging—Act of one throwing an animal, generally mature, by twisting its head and neck.

Bullock—English term for fat steer.

Cake—The mass resulting from the pressing of seeds, meat, or fish in order to remove oils, fats, or other liquids.

Calf—Young animal of cattle species; usually under one year of age.

Cancer eye—Cancerous growth on or around the eye lid.

Carcass—The dressed body of a meat animal, the usual items of offal having been removed.

Casting the withers or eversion of the womb—The womb turns inside out and protrudes from the vagina as a result of excessive straining after calving; sometimes before parturition.

Castrate—Removal of the testicles, or male reproductive organs, from a male animal.

Cattle—More than one of bovine species.

Chute—Equipment constructed for ease in handling cattle, such as a narrow set of panels used for loading.

Cleaned, Cleaning—Removal of material by such methods as scalping, aspirating, magnetic separation, or any other method.

Clip—To shear the hair of an animal, especially around head, udder, etc.

Close Breeding—Inbreeding, such as breeding sire to daughter, son to dam, brother to full sister.

Cod—Scrotum of steer with its content of fat.

Colostrum—The thick viscous milk produced by the dam during the first week of lactation following parturition.

Concentrate—Feed stuffs, such as grains, that are low in fiber and high in total digestible nutrients.

Condition—Degree of fatness in meat animals.

Conformation—Refers to the shape and design of an animal.

Constitution—General bodily vigor indicated by rugged bone, large heart girth, and roomy middle.

Continuous growth—Refers to continuous or uninterrupted development of growing animals as a result of proper nutrition. (Calif. Bull. 688)

Cow—Mature female of cattle species.

Cow hocked—Hocks turn in to each other and often rub together when animal walks.

Creep fed—Refers to animals (generally calves) that are given extra feed by means of small openings in panels that permit only smaller sized animals to enter.

Crest—Back of neck; most pronounced on mature bulls.

Crops—The part of a beef animal just behind the upper half of the shoulders. It extends from the top line about half way down the side.

Crossbred—Animal resulting from mating of purebreds of the same species but of different breeds.

Cubes—Bite-size (1½-inch square) pellets made of processed roughage (alfalfa) having a moisture content of approximately 12 to 13 per cent.

Cud—Regurgitated feed or bolus from stomach.

Cull—Animal taken out of herd because it is below herd standards.

Curry comb—Comb made of several circular blades.

Cutting—Refers to removal of testicles by castration. Also separation of one or more animals from a herd.

Dam—Mother of a calf.

Dehorn—Removal of horns or removal of horn buds so as to prevent any horn growth.

Dewclaw—Rudimentary toes. There are two at the rear and above each pastern joint.

Dewlap—Loose skin found on brisket and neck of some cattle.

Disposition or temperament—A tendency to act in a certain way.

Doctor—One who performs doctoring on animals.

Doctoring—Surgery performed on show animal to hide defects.

Dogie—Slang name for young range cattle.

Dressing per cent—The percentage which the weight of the chilled carcass is of the live weight.

$$\frac{\text{dressed weight}}{\text{live weight}} \times 100 = \text{dressing per cent.}$$

Dry cow—Cow that is not lactating.

Dry weight—Weight of cattle after water is withheld 12 hours.

Dual purpose—Cattle bred and used both for beef and for milk production.

Dwarf—In beef cattle, small abnormal cattle, generally calves, recognized by short broad heads, a bulging forehead, stunted growth, heavy breathing and protruding lower jaw. It is caused by a hereditary factor.

Early maturity—(1) Reaches full development at an early age, (2) reaches market size and finish quickly, (3) tendency to grow and fatten at the same time.

Ear marking—Same as below. See Fig. D.2.

Ear notching—Slits or perforations made in animal's ears for identification purposes. See Fig. D.2.

Ear tagging—Act of putting a tag in ear for purposes of identification.

Embryo transplant—Removing the fertilized egg from one animal (dam) and implanting it into the uterus of another female to develop and produce a live offspring. Thus in cattle a number of calves from

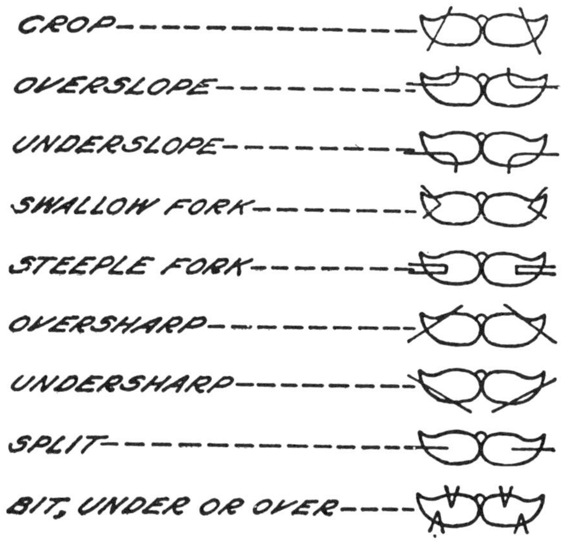

Courtesy, B.A.I., U.S.D.A.

Fig. D.2—Common earmarks of cattle.

a superior animal could be produced in a single year.

EBV—Estimated Breeding Value.

EPD (Estimated Progeny Difference)—The EPD ratio is an estimate of the percentage of the progeny difference from the overall test herd mean that is genetic. For example, if a bull's progeny was 5% above the test herd average and 70% of this difference was genetic, this bull would have an EPD ratio of 103.5 (5 × .7 = 3.5 and 3.5 + 100 = 103.5). The estimated progeny difference is equal to one-half of the breeding value estimate of the bull. In others words, the estimated progeny difference is the amount of extra performance to be expected from future progeny from this bull compared to that expected from future progeny of the average bull in the same test or bred to the same cows.

Fecundity—Ability to produce eggs or sperm regularly.

Feeder—Cattle, often steers, carrying enough age and flesh so as to be ready to place in the feed lot and be grain fattened.

Femininity—Possessing well-developed secondary female sex characters, such as refinement of head and neck and udder development.

Fertile—Able to reproduce regularly.

Fill—The amount of feed and water in an animal.

Finish—Refers to fatness; highly finished means very fat.

Fitting—Refers to the proper feeding, grooming, and handling of an animal so as to make the animal appear at its best.

Flabby—Fleshing in beef cattle that lacks firmness.

Foot—Horny box and its contents; not entire leg.

Freemartin—A sterile heifer twin born with a bull. A

few heifers twinned with bulls are not freemartins; that is, they are able to reproduce.

Full weight—Weight with no special preparation; as is.

Gobby—Lumpy in fleshing; extremely poor distribution of fat.

Grade—An animal not eligible for registration, but one of its parents may be purebred.

Grades of beef—Prime, Choice, Good, Standard, Commercial, Utility, Cutter, Canner (per U.S.D.A.).

Grooming—Brushing or combing animals so as to exercise their skin and improve appearance.

Grubs—Larvae of heel fly found under the skin on backs of cattle.

Heifer—A female of the cattle species usually under three years of age; usually applied to those that have not produced a calf.

Hei-Gro—A small plastic device that is inserted into a heifer's vagina causing her to gain faster and more efficiently.

High grade—An animal possessing 87½ per cent or more of pure breeding, but never 100 per cent.

Hiplock—A condition at parturition whereby the hips of the calf cannot get past the pelvis of the cow so the hips are said to "lock."

Horn weight—Weight used to attach to horn so as to properly shape it.

Hybrid—The offspring obtained by mating two different species.

Hybrid vigor—Increased vigor that accompanies matings which result in hybrids.

Impaction—Over-filling of one or more compartments of the stomach with dry feed.

Inbreeding—Broadly speaking, the mating of related animals. Most breeders describe the mating of closely related animals as inbreeding.

Lactation—Period in which an animal is producing milk.

Leggy—Legs too long, body too high from ground.

Leppy—Orphan calf.

Line breeding—Inbreeding that results when the blood of some animal is introduced more than once, such as grandsire to granddaughter.

Longevity—Refers to life span of an animal, especially as to long-lived animals.

Long fed—Feeding period 120-180 days or longer.

Long yearling—Cattle between 18 months and 2 years of age.

Lousy—Animals infested with lice.

Lumpy jaw—Enlarged jaw caused by a foreign object or infection.

Marbling—Distribution of fat in irregular streaks in lean meat. Gives meat the appearance of marble.

Margin (*spread*)—Difference between purchase price and selling price.

Marking—Overall term used to describe handling process of cattle or calves such as branding, castration, ear notching, etc. Ear marks put on calves for identification purposes are illustrated above.

Masculinity—Possession of well-developed secondary male characteristics in head, neck, shoulders, and crest.

Mature—Age reached at which there is little increase in size of animal and after animal is capable of or has produced its first offspring.

Maverick—An unbranded animal on the range.

MPPA—Most probable producing ability.

Muley—Naturally hornless or polled.

Natural fleshing—Lean meat or muscle.

Nick (*niche*)—Results of certain matings which produce an animal of high order of excellence; sometimes from mediocre parents.

Nurse cow—Lactating cows which are used to nurse calves other than their own.

Off feed—Cattle that stop eating or eat very little; most often occurs with fattening cattle.

Offal—All organs or tissues removed from inside of the carcass in slaughtering.

On the prod—Angry cow; generally cross enough to pursue humans.

On the road—Exhibition cattle traveling the show circuit.

Orejana—Orphan calf.

Outcross—The introduction into a herd which is more or less related of blood from some outside and unrelated source but of the same breed.

Overshot—Applies to jaw of cattle when upper jaw protrudes over lower jaw.

Parturition—The process of the female giving birth to young.

Patchy—Having lumps of exterior fat which prevent the smooth finish desired.

Paunchy—Applied to an animal which carries too much belly.

REPRODUCTION CYCLE OF DOMESTIC ANIMALS

Animal	Occurrence of Cycles	Average Duration of Cycles (Days)	Duration of Heat	Time of Ovulation	Duration of Pregnancy (Days)
Cow	All year	21	8-24 hours	First day after end of heat	275-290
Ewe	Fall—varies with breed and climate	18	1-3 days	At end of heat	145-155
Mare	Spring Feb.-July	22	2-6 days	Near end of heat	325-345
Sow	All year	21	1-4 days	Toward end of heat	112-120

Pedigree—A written statement giving the record of an animal's ancestry.

Penectomy—Surgical alteration of the location of the external male organ, leaving all tissues intact but making copulation impossible.

Plain headed—Undesirable type of head.

Poll—Top of head.

Polled—Naturally hornless; a muley.

Prepotency—The ability of an animal to transmit its own qualities to its offspring.

Pressure sensitive pads.—Dye-filled pad. When glued to a cow's back, it ruptures under the weight of another animal mounting her.

Prolapsed vagina—Inverted womb.

Prolific—The ability to produce regularly and numerously.

Proud flesh—Excess flesh growing around wound.

Purebred—An animal of pure breeding, eligible for registration in its breed association.

Quality—Fineness of texture; freedom from coarseness; generally used to indicate superior breeding or merit.

Ralgro—A feed additive, generally used as an implant, that improves rate of gain and feed efficiency. Chemical compound is Zeranol. It is an anabolic agent for growth promotion.

Range—Area on which natural feed is produced and harvested by beef cattle.

Rangy—Too much length of body associated with too much length of leg.

Ratio—A device used to indicate an animal's ranking in relation to rest of the herd.

$$\frac{\text{individual record}}{\text{average of animals in group}} \times 100$$

Ration—Strictly speaking it is the amount of feed allowed an animal in one day or 24 hours.

Registered—Purebred animals whose pedigree is recorded in the breed registry.

Replacement—Animals selected to keep in the breeding herd.

Reproduction—The process, sexual or asexual, by which plants or animals produce new individuals or offspring.

Ringer—Illegal substitute for a show entry.

Roman nose—The profile of the face as viewed from side is convex.

Roughage—Coarse feeds, such as hay, that are relatively high in fiber and low in total digestible nutrients.

Rugged—Big, strong, and sturdy.

Rumensin (monesin sodium)—A feed additive (not a hormone) which changes rumen digestion so cattle get more energy from feed.

Scale—Size.

Scotch comb—Straight or single-blade comb.

Scours—Diarrhea or loose running feces.

Scrotal circumference—The distance around (circumference in centimeters) the scrotal sac at its largest diameter. Yearling bulls less than 30 cm. are viewed with suspect relative to their semen producing ability.

Scrub—A very inferior animal in breeding and conformation.

Scurs—Small, imperfectly formed horns not attached to the skull.

Self feeding—Equipment designed to supply feed to animals so they can eat at will.

Selling on the rail—Selling an animal on a dressed carcass basis, as it hangs in the packing house cooler.

Shank—In the live meat animal, it is the part of the leg between the knee and ankle and between the hock and ankle.

Short fed—Feeding period 90 to 120 days or less.

Short two—Cattle just under two years of age.

Shoulder vein—The line formed by junction of side of neck and shoulder.

Show box (tack box)—Container used to keep all show paraphernalia.

Show string—Group of cattle selected for exhibition purposes.

Shrinkage, "shrink"—Weight lost by cattle during transit or other handling processes.

Shy breeder—A female who tends to be difficult to get with calf.

Sickle hocked—Hock joint is bent so that animal stands with hind feet forward and under the body.

Sire—Male animal that is the paternal parent of a calf.

Spayed heifer—A heifer whose ovaries have been removed by an operation.

Spotter bull—A vasectomized bull used to determine which females are in heat.

Springing—Bred females that show considerable evidence of approaching parturition.

Squeeze—Equipment similar to a chute except that one side is hinged to permit more thorough restraint of animal while operations, such as branding, are performed.

Stag—A male which was castrated after reaching breeding age. Secondary sexual characters are more or less developed.

Steer—Male of cattle species castrated at an early age.

Stocker—Cattle, any age or sex, but generally younger, thinner animals, used for wintering on roughage with little or no grain.

Stocks—Equipment used in restraining cattle; generally consists of roller and sling designed to lift animal so it cannot lie down or throw itself while minor operations are performed.

Parts of Steer

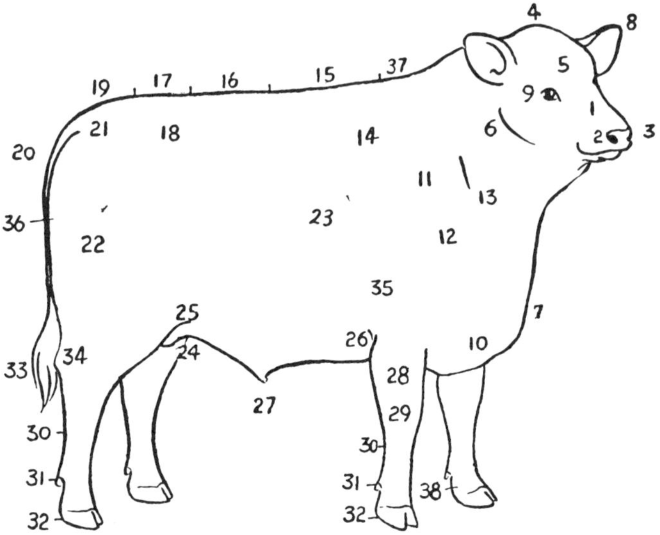

Fig. D.3

1. Face
2. Nostril
3. Muzzle
4. Poll
5. Forehead
6. Neck
7. Dewlap
8. Ear
9. Eye
10. Brisket
11. Shoulder
12. Point of shoulder
13. Shoulder vein
14. Crops
15. Back
16. Loin
17. Loin end
18. Hooks
19. Tailhead
20. Tail
21. Rump
22. Thigh
23. Ribs
24. Cod
25. Hindflank
26. Foreflank
27. Pizzle
28. Arm
29. Knee
30. Shank
31. Dewclaw
32. Foot
33. Switch
34. Hock
35. Forerib
36. Twist
37. Top of shoulder
38. Pastern

Stylish and attractive—Possessing a pleasing conformation coupled with animation and gracefulness of carriage.

Supplement—Feeds, such as many of the high protein feeds, that are used to balance or improve existing rations.

Tail-tie—A tie on the tailhead caused by hide adhering to bone.

Tattoo—Permanent identification of animals produced by putting indelible ink under the skin; generally put in the ears of animals while they are young.

Teaser—Animals used to seek out females that are in heat.

Terminal cross—The end generation of a cross breeding system. None of these individuals is kept for breeding.

Thigh—The outside of the hind leg between the rump and hock. Should be wide and bulging, carrying down as close to the hock as possible.

Throw—Restraining an animal by use of a rope or other force so as to render it helpless in order that certain operations may be performed.

Tie—A depression or dimple on the back of cattle caused by adhesion of hide to vertebrae.

Twist—Fleshing between the hind legs. Should be full and deep.

Type—The general desired form of an animal which adapts it for a particular purpose.

Typy—Possessing the desired form.

Undershot—Applies to jaw of cattle when lower jaw protrudes further than upper jaw.

Vaccination or shot—Introduction of antibiotics into the body of an animal.

Vasectomy—Surgical removal of a section of the vas deferens preventing sperm discharge.

Vasectomized—A condition whereby a portion of the sperm cords of a male has been cut or removed so that he is sterile yet will go through the act of breeding.

Wasty—Applied to carcass which has too much fat; requires excessive trimming; may also be applied to paunchy live animals.

WDA (Weight-per-Day-of-Age)—The WDA includes

birth weight and is calculated from date of birth to weigh day. It is the actual lifetime average daily gain and does not reflect any adjustment factors.

Wean—Removing the calf from its dam so that it can no longer nurse.

Weaner—Calf just taken off the dam.

Wooden tongue—Disease in which tongue becomes stiff.

Yearling—Animal approximately one year old.

Use Proper Terminology—It is a desirable practice to identify the parts of an animal by correct terms. See Fig. D.3.

INDEX